AF782628

ENERGY POLICIES, POLITICS AND PRICES

Additional books in this series can be found on Nova's website under the Series tab.

Additional E-books in this series can be found on Nova's website under the E-book tab.

ENERGY SCIENCE, ENGINEERING AND TECHNOLOGY

Additional books in this series can be found on Nova's website under the Series tab.

Additional E-books in this series can be found on Nova's website under the E-book tab.

ENERGY POLICIES, POLITICS AND PRICES

PHOTOVOLTAICS

LOCAL INDUSTRY DEVELOPMENT, INSTALLED COST TRENDS AND MINERAL COMMODITIES USED

ERIC FALCHUK
AND
CAROL WOODLEE
EDITORS

Nova Science Publishers, Inc.
New York

Copyright © 2012 by Nova Science Publishers, Inc.

All rights reserved. No part of this book may be reproduced, stored in a retrieval system or transmitted in any form or by any means: electronic, electrostatic, magnetic, tape, mechanical photocopying, recording or otherwise without the written permission of the Publisher.

For permission to use material from this book please contact us:
Telephone 631-231-7269; Fax 631-231-8175
Web Site: http://www.novapublishers.com

NOTICE TO THE READER

The Publisher has taken reasonable care in the preparation of this book, but makes no expressed or implied warranty of any kind and assumes no responsibility for any errors or omissions. No liability is assumed for incidental or consequential damages in connection with or arising out of information contained in this book. The Publisher shall not be liable for any special, consequential, or exemplary damages resulting, in whole or in part, from the readers' use of, or reliance upon, this material. Any parts of this book based on government reports are so indicated and copyright is claimed for those parts to the extent applicable to compilations of such works.

Independent verification should be sought for any data, advice or recommendations contained in this book. In addition, no responsibility is assumed by the publisher for any injury and/or damage to persons or property arising from any methods, products, instructions, ideas or otherwise contained in this publication.

This publication is designed to provide accurate and authoritative information with regard to the subject matter covered herein. It is sold with the clear understanding that the Publisher is not engaged in rendering legal or any other professional services. If legal or any other expert assistance is required, the services of a competent person should be sought. FROM A DECLARATION OF PARTICIPANTS JOINTLY ADOPTED BY A COMMITTEE OF THE AMERICAN BAR ASSOCIATION AND A COMMITTEE OF PUBLISHERS.

Additional color graphics may be available in the e-book version of this book.

Library of Congress Cataloging-in-Publication Data

ISBN: 978-1-61942-045-8

Published by Nova Science Publishers, Inc. † New York

CONTENTS

PREFACE

The U.S. photovoltaic (PV) industry is forecast to grow in both the short term and the long term. It represents a significant opportunity for economic development and job creation in communities throughout the United States. The PV industry is best regarded as two major sectors: PV products (a sector of businesses that manufacture raw materials, PV modules and other components) and PV energy generation (a sector of services centered on the purchase of components and the installation of PV systems). This book provides background on the PV industry with a focus on helping communities evaluate opportunities in the photovoltaic industry and develop a strategic approach appropriate to a specific community.

Chapter 1- The U.S. photovoltaic (PV) industry is forecast to grow in both the short term and the long term. It represents a significant opportunity for economic development and job creation in communities throughout the United States.

For the purposes of designing economic development efforts, the PV industry is best regarded as two major sectors: PV products (a sector of businesses that manufacture raw materials, PV modules, and other components) and PV energy generation (a sector of services centered on the purchase of components and the installation of PV systems).

The differences between these two sectors are important because they each have unique economic development strategies.

Chapter 2- As the deployment of grid-connected solar photovoltaic (PV) systems has increased, so too has the desire to track the installed cost of these systems over time and by location, customer type, system characteristics, and component. This report helps to fill this need by summarizing trends in the installed cost[1] of grid-connected PV systems in the United States from 1998 through 2009 (updating two previous reports with data through 2007 and 2008, respectively), and providing preliminary cost trends for systems installed in 2010.[2] The analysis is based on installed cost data for approximately 78,000 residential and non-residential PV systems, totaling 874 megawatts (MW) and representing 70% of all grid-connected PV capacity installed in the United States through 2009.

Chapter 3- Metals such as cadmium, gallium, germanium, indium, selenium, and tellurium are important mineral materials used in current photovoltaic cell technology. Most of the world's primary supply of these mineral commodities is recovered as byproducts from ores processed mainly for the purpose of producing aluminum, copper, lead, and zinc and, in the case of germanium, from ash derived from burning coal for the production of energy. In some cases, the refined metals originate in the countries in which they are mined, while in other cases, intermediate products containing the metals and refined metals are recovered by processers located in other countries. Assuming that the modest rate of market penetration of electricity from photovoltaic cells and competing markets increases demand for cadmium, gallium, and germanium, the resulting new demand can be met in relatively quick order with the expansion of existing recovery circuits or the addition of new circuits. It is very unlikely that, under most scenarios, the need would arise to mine more bauxite, copper, and zinc ore or to burn more coal to satisfy anticipated material requirements to meet the photovoltaic-generated demand for these mineral materials. For other metals, such as indium, selenium, and tellurium, targeted mineral exploration and improvements in metallurgical recoveries are likely to be needed to meet demand requirements. Substitution for these materials in nonphotovoltaic applications could also "free up" supply. Conversely, however, other new applications can produce supply constraints or increase metal prices and place pressure on their use in photovoltaic technology.

In: Photovoltaic
Editors: Eric Falchuk and Carol Woodlee

ISBN: 978-161942-045-8
© 2012 Nova Science Publishers, Inc.

Chapter 1

SOLAR PHOTOVOLTAIC ECONOMIC DEVELOPMENT: BUILDING AND GROWING A LOCAL PV INDUSTRY*

U.S. Department of Energy's Solar America Communities

ABBREVIATIONS AND ACRONYMS

>	greater than
<	less than
$/W	dollars per watt
AC	alternating current
ARRA	American Recovery and Reinvestment Act
BETC	Business Energy Tax Credit
BOS	balance of system
CAPEX	capital expenditures
CEMF	Clean Energy Manufacturing Fund
CdTe	cadmium telluride
CIGS	copper indium gallium diselenide
CIS	copper indium diselenide
c-Si	crystalline silicon
CPV	concentrating photovoltaic
CSLP	
DC	direct current
DOE	U.S. Department of Energy
EPC	engineering, procurement, and construction
F&E	franchise and excise (tax)
FIT	feed-in tariff

* This is an edited, reformatted and augmented version of The United States Department of Energy's Solar America Communities publication, prepared for the Unite States Department of Energy, Solar Energy Technologies Program by CH2MHill, dated August 2011.

FTE	full-time equivalent
GRU	Gainsville Regional Utilities
GW	gigawatt
IREC	Interstate Renewable Energy Council
JEDI	Jobs and Economic Development Impact
JTIP	Job Training Incentive Program
KEPCO	Korea Electric Power Corpoation
kg	kilogram(s)
kW	kilowatt(s)
kWh	kilowatt-hour(s)
kWh/kW	kilowatt-hour(s) per kilowatt
MW	megawatt(s)
NREL	National Renewable Energy Laboratory
NYSERDA	New York State Energy Research and Development Authority
OPEX	operating expenditures
OEWD	Office of Economic and Workforce Development
PACE	Property Assessed Clean Energy
PBI	production-based incentive
PPA	power purchase agreement
PV	photovoltaic(s)
R&D	research and development
REC	renewable energy certificate
RETA	Renewable Energy Transmission Authority
RFP	request for proposals
RPS	renewable portfolio standard
SETP	Solar Energies Technologies Program
SIC	Standard Industrial Classification
W	watt(s)
W_p	(peak) watt(s)

ABOUT THIS DOCUMENT

Purpose

This report is intended to be an introductory guide for local economic development offices to set informed recruitment targets for renewable energy. It was developed to help communities evaluate opportunities in the photovoltaic (PV) industry and develop a strategic approach appropriate to a specific community. The document provides the information for communities to:

1) Build an understanding of the various sectors of the PV industry, metrics, differentiation, and trends in each segment

2) Build an understanding of site selection processes and criteria for each industry segment, and the potential economic benefits of each segment
3) Begin to evaluate the city's strengths, weaknesses, and potential to change within the context of the industry segments
4) Understand tools and programs used to encourage investment and economic development. Use these tools to reinforce strengths, improve weaknesses, and market to the PV industry over the long term.

This report summarizes the PV industry, but further information and reading may be important to city-specific efforts. Appendix A describes additional resources.

Organization

The PV industry is an umbrella category that represents many diverse business models. In order to discuss and review the whole industry effectively, this report groups the industry into two major sectors: Section 1 discusses the PV products and energy generation sectors, with a particular focus on details relevant to economic development entities. This grouping splits the industry according to basic business models and allows some comparisons between similar business activities. Section 2 describes the employment opportunities available through the PV products and energy generation sectors, and describes economic development strategies targeted at attracting these employers. Section 3 provides case studies of economic development programs and strategies successfully employed by states and local jurisdictions in the United States. Appendix A provides a list of recommended resources, while Appendices B and C provide additional details on the value chain and business structures employed in both sectors.

Methodology

Financial analyses were completed by a financial expert at CH2M HILL with 28 years of professional experience, including the creation of financial models for PV manufacturers, solar farms, and financial institutions working with the PV industry.

Primary-source research comes from communication with researchers at the U.S. Department of Energy (DOE) and market research firms. Case studies were created through communication with local economic development experts and other sources, as cited.

Secondary-source research comes from studies conducted by the DOE, local and national economic development agencies, and market research firms, as cited.

Executive Summary

The U.S. photovoltaic (PV) industry is forecast to grow in both the short term and the long term. It represents a significant opportunity for economic development and job creation in communities throughout the United States.

For the purposes of designing economic development efforts, the PV industry is best regarded as two major sectors: PV products (a sector of businesses that manufacture raw materials, PV modules, and other components) and PV energy generation (a sector of services centered on the purchase of components and the installation of PV systems).

The differences between these two sectors are important because they each have unique economic development strategies.

The PV Products Sector

Within the PV products sector, there are two segments. The dominant segment is based on crystalline silicon (c-Si) modules. There is also a growing segment based on thin-film technology modules. Figure 1 shows the value chain elements for these two segments.

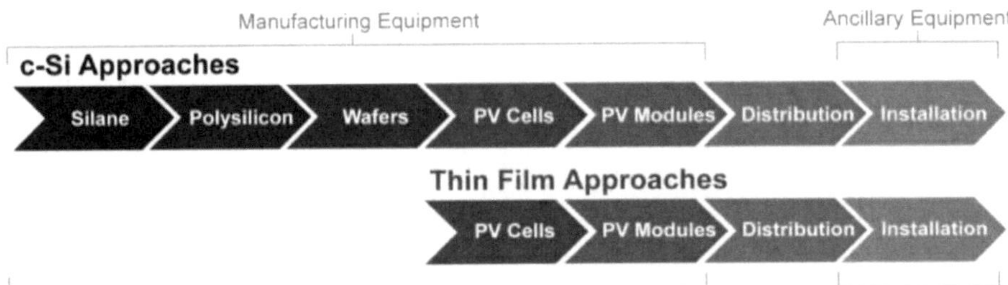

Source: O'Rourke et al. 2009.

Figure 1. PV products sector value chain.

Metrics, Markets, and Trends

Companies in the PV products sector generally share the following major metrics:

- Product performance (power output in watts)
- Efficiency (percentage of sunlight or energy that is converted into usable electricity)
- Product cost (dollars per watt)
- Reliability (product life and performance covered by warranties).

Typically[1], markets for these PV products are differentiated by their scale, as follows:

- Residential—less than (<) 10 kilowatts (kW)
- Commercial—10 kW to 1 megawatt (MW)
- Utility—greater than (>) 1 MW.

The PV products sector is also characterized by several general trends:

- A shake-out of manufacturing companies driven by the economic downturn
- Integration of companies up and down the supply chain to deal with the boom-and-bust cycles in raw materials and with the competition from international markets
- Diversification of products driven by the expansion of PV applications.

PV Products Sector Site Selection Criteria

Understanding how companies in the PV products sector evaluate a location is critical to understanding an area's competitive position and developing strategies to improve that position. For the purposes of this report, the products sector has been divided into four industry groups:

- Raw materials (such as polysilicon, glass, metal products, and gases) used in the production of panels
- Manufacture of wafers and cells that are integrated into panels
- Assembly of components into panels
- Products (such as inverters, racking, and wiring) that are integrated into and support panels.

Table 1 presents a summary matrix of the PV products sector requirements by industry segment and their relative importance. Section 1.1 and Appendix B discuss these requirements.

Table 1. Matrix of PV Products Sector Site Selection Criteria

	PV Products			
Industry Criteria	Raw Materials	Wafer/Cell	Assembly	Other Supply Chain
Land and Buildings	2	2	2	2
Utilities	1	1	3	3
Operating Cost Environment	1	1	2	2
Schedule Considerations	1	1	2	3
Transportation	2	2	1	2
Proximity to End Market	2	2	1	2
Technology Support, Higher Education and Research	2	1	3	3
Workforce and Training	1	1	2	2
Public/Private Support and Policies	2	1	1	2
Incentives	1	1	2	2

Legend: Critical (1) Somewhat Important (2) Less Important (3)

The PV Energy Generation Sector

Figure 2 shows the various segments within the PV energy generation sector. Many companies span multiple segments, and third-party companies (such as financial institutions) may operate across various segments as well.

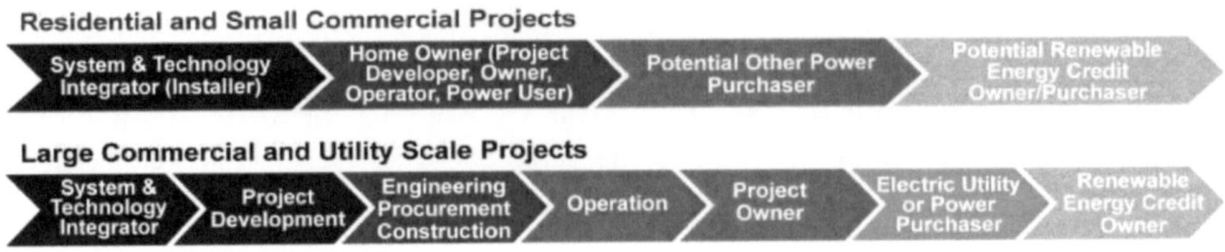

Figure 2. PV energy generation sector value chain.

Markets and Trends

PV energy generation markets are typically characterized in terms of residential, commercial, and utility scale. PV energy generation services can also be analyzed in terms of the following:

- Project scope. From locally labor-intensive segments, like system integration and installation, to low-labor-intensity segments like project development
- Project integration. Energy generation services that are often bundled together
- Project risk. From existing and proven models of ownership to new legal or financial structures.

The PV energy generation sector can be characterized by several general trends:

- An increasing separation between owners and hosts, exemplified by the third-party power purchase agreement project structure
- An increase in the participation of electric utilities
- Increased integration of services into single contracts
- Diversification and specialization of services within the PV energy generation sector as increasingly complex project opportunities emerge.

PV Energy Generation Sector Site Selection Criteria

The location factors and criteria considered by PV energy generation companies are relatively concise and straightforward (compared with PV products companies). They are driven primarily by market proximity, as well as by the criteria presented in Table 2.

Table 2. Matrix of PV Energy Generation Market Criteria

	Energy Generation	
Industry Criteria	Residential / Commercial	Utility Scale
Solar Resource	2	1
Cost of Electricity, Utility Structure and Attitudes	1	1
Public / Private Support and Policies	1	1
Workforce and Training	3	3

Legend: Critical (1) Somewhat Important (2) Less Important (3)

Employment Opportunities and Economic Development Strategies

While PV products sector jobs vary by market segment, they are generally stable, highly skilled, and long term. Because of the large investments required, and the high level of competition for these facilities, recruiting or growing a large PV products sector company is complex and can require significant investment, planning, and long-term commitment. Not all communities have the same opportunities, but those that have been successful have capitalized on economic assets while providing incentives to compensate for shortcomings.

PV energy generation jobs tend to be more variable. Installation labor often represents the largest opportunity for economic development and job creation. PV energy generation jobs are dependent on project durations and are often short term. They can, however, be more permanent if the community commits to the long-term development of energy generation projects. Because of local conditions like solar resources and local utility structures, building an attractive market for PV energy generation companies usually requires strong public policy support and investment incentives.

CONCLUSION

Many U.S. communities have some opportunity to develop a segment of the solar PV industry. While not every community can recruit a large factory or install megawatts per year of distributed PV on local homes, many have high potential in some segments and can invest in and recruit a part of the PV industry—thereby creating jobs and bolstering the local economy.

Precursors to successful PV industry recruitment include the following:

- Understand the segments of the industry and what is required to operate successfully in an area
- Make a realistic evaluation of how well a community meets the requirements of various segments
- Target a market segment(s) that best meets community economic development goals and whose siting requirements are satisfied by the community's economic capabilities
- Develop strategies to strengthen advantages and overcome disadvantages
- Market opportunities to target segments.

1. PHOTOVOLTAIC INDUSTRY BASICS

While most industry analysts agree that the future outlook and forecasted economic impact of photovoltaics (PV) at the national scale is very positive, it is important to distinguish the various sectors of the U.S. PV industry, each of which has different needs, drivers, and success metrics. This report focuses on two major PV sectors: PV products and PV energy generation.

These two PV industry sectors are fundamentally different economic development targets. Appendices B and C describe the business and manufacturing processes utilized in each of these sectors in more depth, highlighting those details relevant for economic developers.

1.1. Photovoltaic Products

For the purposes of this report, the PV products sector is defined as businesses that design, manufacture, or distribute PV equipment, including PV panel subcomponents and consumables, PV panels, inverters, racking, and other components (called the balance of systems or BOS). At its core, it is a sector defined by manufacturing.

Growth in manufacturing of PV products is illustrated in Figure 3, which shows the manufacturing capacity of U.S. PV product companies measured in megawatts (MW) of PV cells.

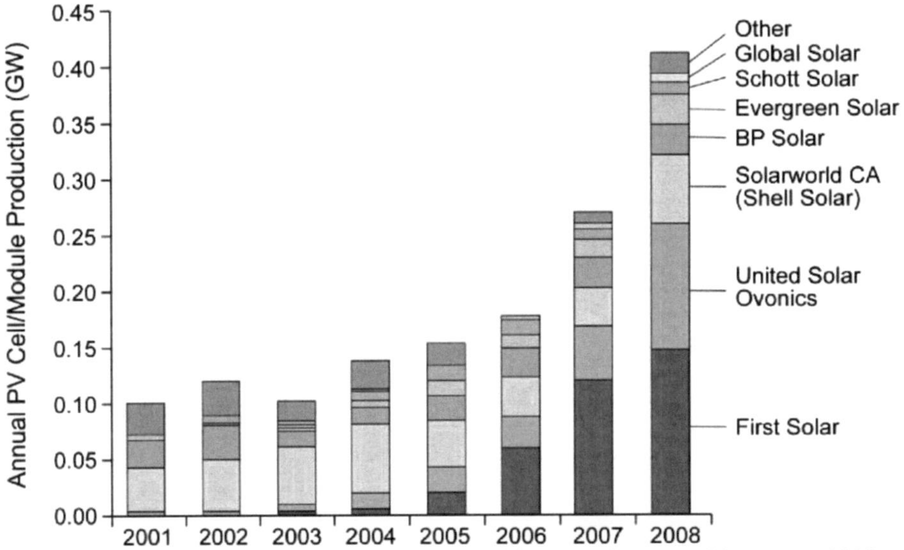

Source: U.S. Department of Energy Office of Energy Efficiency and Renewable Energy 2010.

Figure 3. U.S. annual PV cell/module production by U.S. manufacturers.

The major businesses in the PV products sector manufacture or fabricate the following products:

- Polysilicon
- Solar glass
- Bulk chemicals and gases
- Manufacturing equipment and parts
- Junction boxes and connectors
- Other cell and module materials (such as films, string, and silver paste)
- Module frames

- Silicon wafers
- PV cells
- PV modules
- Inverters
- Wiring, conduit, and connectors
- Transformers and switchgear
- Racking
- Trackers
- Sensors, controls, and actuators
- Anchors and ballasts
- Optical components.

These products are commodity-like in nature. For example, while brand, country of origin, or other factors can influence buying decisions, these influences are generally minor—and shrinking. Exceptions to this trend are "Buy American" or local content provisions in incentives, and specific rules for certain potentially hazardous materials (such as lead or cadmium) in Japan and the European Union.

The focal point of the PV products sector is the functional unit of a PV system, the PV module. Typical system components and their estimated contribution to the installed cost of a typical PV system, differentiated by market segment, are illustrated in Figure 4.

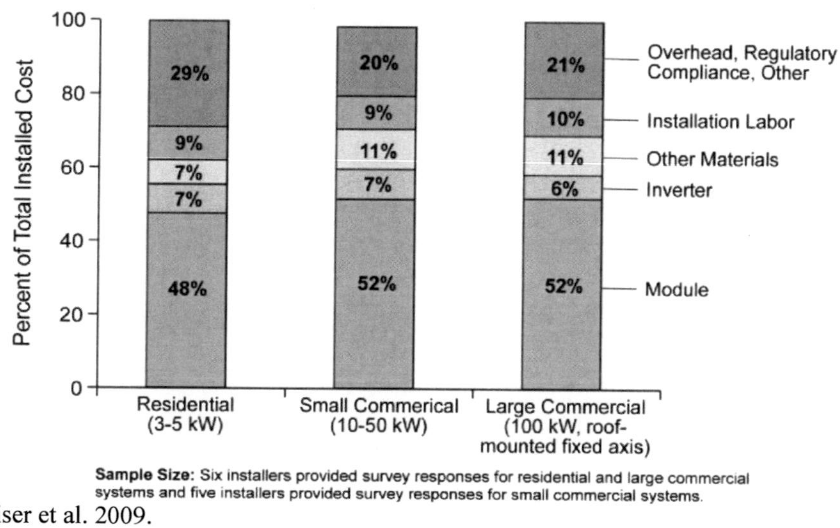

Source: Wiser et al. 2009.

Figure 4. PV system component contribution to total system costs.

The products listed in Figure 4 are produced by the supply chain shown in Figure 5. The depicted supply chain divides the PV products industry into two fundamental manufacturing strategies that use different materials and different technologies to make PV modules:

- *Crystalline silicon (c-Si).* This dominant technology based on bulk silicon, uses proven manufacturing technology and well-understood materials, has a history of improvement, and has a strong road map for modest future improvements.

- *Thin film.* This newer technology is based on very thin layers of other semiconductor materials, deposited on glass or another substrate. It uses a variety of manufacturing technologies and core materials, each with its own challenges. The history of thin film is shorter than c-Si, and its manufacturing process development is of somewhat higher risk. However, thin film has more room than c-Si for cost and performance improvements before approaching theoretical maximums.

Further discussion of the PV products supply chain is included in Appendix B.

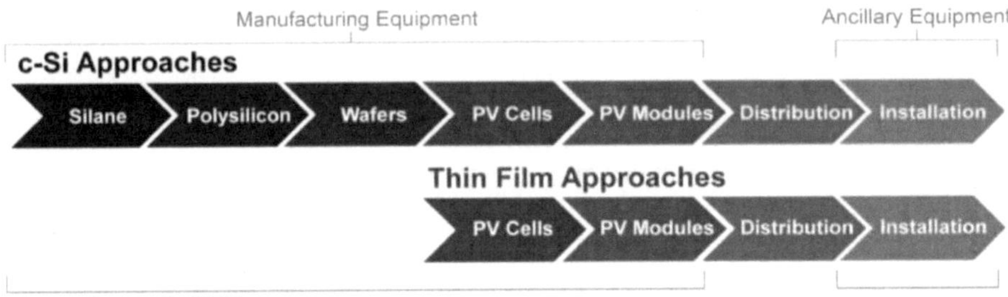

Source: O'Rourke et al. 2009.

Figure 5. PV products sector supply chain.

1.1.1. Sector Metrics

PV product performance is typically measured by peak watts (Wp), the peak power a PV product is capable of generating or handling under ideal installation conditions, by percent efficiency and reliability:

- Peak watts is a measurement of power, and is the metric by which PV products are sold, often dollars per watt ($/W). This is often confused with kilowatt-hours (kWh), a measurement of the energy generated that depends on a variety of system and geographic factors.
- Efficiency is the percentage of solar energy falling on the PV device that is converted to usable electricity. Efficiencies are quoted for many individual PV products, with system efficiency always lower than the efficiency of any one component. Typical commercially available panel efficiencies (not including concentrating photovoltaic, or CPV, systems) range from 6% to 25%, and system efficiencies range from 5% to 20%. System efficiencies are often guaranteed by the manufacturer.
- Reliability is typically defined as the usable lifetime of the product under forecast conditions, and its ability to perform predictably. Typical commercially available ranges are 10 to 30 years, with some minor per-year performance degradation typical of most PV systems. These lifetimes are often guaranteed by the manufacturer.

The PV products sector's warranties on their products (and their ability to make good on those warranties) is a variable that many energy generation sector businesses watch closely.

1.1.2. Sector Differentiation

PV products are differentiated according to market scale and cost-per-performance, as described below.

1.1.2.1. Market-Scale Differentiation

PV products are typically intended for one of three markets, defined by their scale: residential, commercial, and utility. Although there is some crossover, these markets are becoming increasingly differentiated. The most significant difference exists between residential/commercial-scale and utility-scale projects (Figure 6).

1.1.2.2. Cost-per-Performance Differentiation

The c-Si strategy targets high performance (lowering cost per watt by increasing efficiency and W_p). The thin-film strategy targets lower costs (including focusing on reducing material usage and manufacturing complexity).

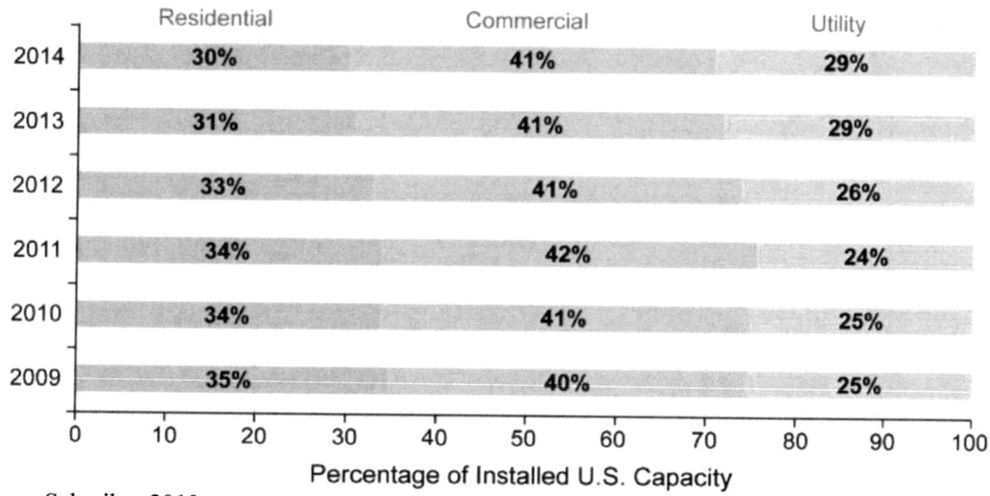

Source: Schreiber 2010.

Figure 6. U.S. market segmentation 2009–2014.

These two product differentiation themes are interrelated; the general trend shows that the smaller the market scale for which the product is intended, the greater the focus on high performance.

1.1.3. Supply Chain—Raw Materials to Finished Products

The PV products sector supply chain involves many discrete manufacturing businesses. These major activities include the manufacture of key raw materials, PV modules, and other supply chain components. While raw materials manufacturing has a high barrier to entry based on technology and economies of scale, downstream manufacturing operations have decreasing barriers and significantly more competition.

Many of the following supply chain steps are often geographically distant, but some can also be integrated at a single site. Locations ideal for such integrated facilities meet the combined site selection criteria listed at the end of this section.

1.1.4. Trends

The PV products sector is growing and changing. The following paragraphs describe some ongoing trends affecting companies throughout the supply chain.

1.1.4.1. Influence of the Economic Downturn

From 2005 through 2008, a lot of investment and energy went into PV manufacturing research and development (R&D). Manufacturing capacity was being built and expanded as fast as construction schedules would allow. Manufacturers expected endless demand. In the last part of 2008, 2009, and into 2010, however, the general economic downturn cut PV demand, and growing inventories led to falling average selling prices. While the long-term outlook is strong, the industry will face many challenges; therefore, investments must have strong economic justifcation.

1.1.4.2. Continued Vertical Integration

While no supply-chain bottlenecks are currently driving vertical integration in the PV products sector, a history of boom-and-bust cycles continues to drive PV products sector companies to seek partners and acquisitions. By integrating vertically and buying or developing upstream and downstream capabilities, companies hope to increase their internal coordination and cost and margin control and reduce their exposure to supply-chain imbalances. If companies cannot buy or develop these vertically integrated capabilities, they are trying to sign long-term contracts and form strong ties within the supply chain.

As a result of the economic downturn, many of the weakest companies simply have failed. These companies' intellectual property, industry scientists and experts, and valuable tools, equipment, facilities, and staff are being acquired by the stronger companies.

1.1.4.3. Differentiation and Diversification

Market maturity is slowly defining various market niches by scale, location, and needs for intelligence, and shaping products that best fit them. Building-integrated photovoltaics is a great example of this trend, as PV products increasingly are being integrated into standard building materials, as well as into other devices, appliances, consumer products, and road signage and lighting.

At the same time, other functions and intelligence are being integrated into PV products, such as modules that optimize output under shading conditions through integrated electronics.

While these innovative market niches are exciting, they are not forecast to be a meaningful portion of the overall PV product market. The market will continue to be driven by on-grid residential, commercial, and utility installations.

1.1.5. Site Selection Criteria

When companies in the PV products sector choose a new place to manufacture products, they evaluate areas based on the following key criteria:

- Land and buildings
- Utility service for electricity, water and wastewater treatment, and natural gas
- Operating cost environment
- Ability to meet production schedules and deadlines

- Transportation
- Proximity to PV end market
- Technology support, higher education, and research
- Workforce and training
- Public and private support and policies
- Incentives.

Informing city-level economic development strategies, communities need to consider how they measure up against these metrics, and what they can do to improve their scores. Table 3 illustrates the relative importance of these site selection criteria to companies in various parts of the PV products sector. Communities can use these criteria to quickly evaluate themselves relative to the needs of the industry segment they are considering targeting. In so doing, communities can identify potential strengths, challenges, and criteria on which they must improve to be successful. Note: the rankings in Table 3 are solely the opinion of the authors. These rankings are a first-order attempt at characterizing the siting criteria utilized in the PV products sector, included here solely for the purposes of informing economic development planning.

1.2. Photovoltaic Energy Generation

For the purposes of this report, the PV energy generation sector is defined as businesses that (a) model, design, or install PV systems, or (b) generate, distribute, sell, and/or provide any guarantees on system performance. At its core, it is a sector defined by technical and construction services, most important of which is the installation of PV systems.

Table 3. Matrix of PV Product Sector Site Selection Criteria

	PV Products			
Industry Criteria	Raw Materials	Wafer/Cell	Assembly	Other Supply Chain
Land and Buildings	2	2	2	2
Utilities	1	1	3	3
Operating Cost Environment	1	1	2	2
Schedule Considerations	1	1	2	3
Transportation	2	2	1	2
Proximity to End Market	2	2	1	2
Technology Support, Higher Education and Research	2	1	3	3
Workforce and Training	1	1	2	2
Public/Private Support and Policies	2	1	1	2
Incentives	1	1	2	2

Legend: Critical (1) Somewhat Important (2) Less Important (3)

Growth in PV energy generation is illustrated in Figures 7 and 8, which highlight the increasing growth of grid-connected PV, as well as the importance of commercial- and utility-scale installations compared with residential installations. Grid-connected PV is the market for PV installations interconnected to the general electrical transmission and distribution system. This is in contrast with off-grid PV markets, which are systems often paired with battery backup systems for rural or remote power. Residential-, commercial-, and utility-scale PV refer to different general markets in PV energy generation defined by their scale.

The PV energy generation sector is a services sector. Major services and businesses in the PV energy generation sector include:

- Array and system design
- Electrical and mechanical contractors
- Power producers
- Engineering, procurement, and construction
- Financial and legal consulting
- Industry organizations and lobbyists
- Logistics and transportation
- Operations and maintenance
- Owner's engineer
- Performance modeling
- Performance monitoring
- Site and civil work
- Site selection
- System finance (including lease and third party)
- Tax investor
- Training organizations
- Transmission and interconnection studies.
-

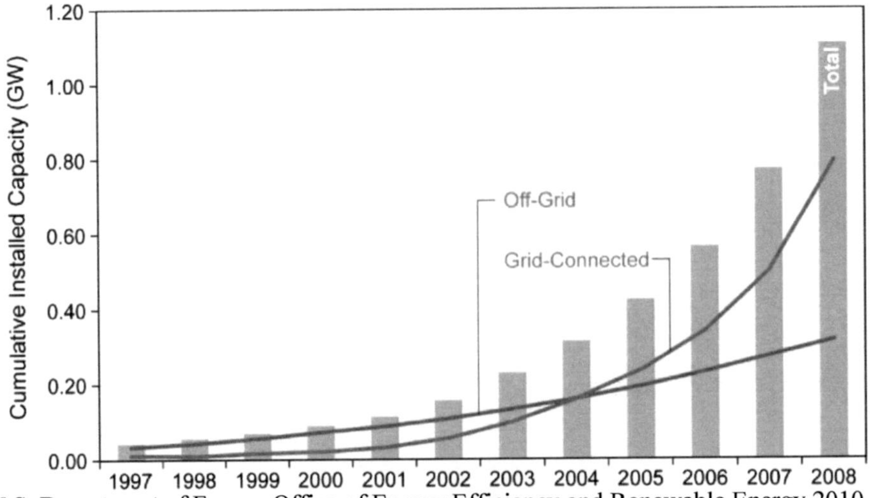

Source: U.S. Department of Energy Office of Energy Efficiency and Renewable Energy 2010.

Figure 7. U.S. cumulative installed PV capacity by interconnection status.

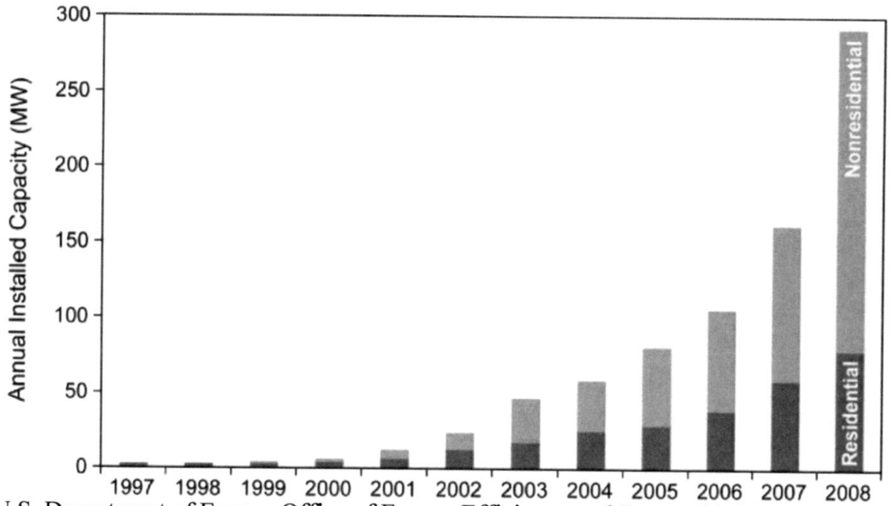

Source: U.S. Department of Energy Office of Energy Efficiency and Renewable Energy 2010.

Figure 8. U.S. annual grid-connected PV capacity by sector.

The focus of PV energy generation activities is the delivery of a kilowatt-hour from a PV system to a customer; however, the range of services in this category is very broad. In general, PV energy generation activities strive to purchase and install PV products, operate them, and make their output available to energy users.

In forming local economic development strategies for this sector, it is important to think in terms of building an attractive market for PV projects (installations), not recruiting other PV energy generation services companies that provide relatively few direct and indirect local jobs, or that can operate distant from project sites. The majority of jobs from the PV energy generation sector are in project installation labor.

Figure 9 illustrates the value chain for the PV energy generation sector. The value chain is divided between residential and small commercial projects, and large-scale commercial or utility projects. Larger scale, more complex projects benefit from increased specialization and more complex agreements, often involving third parties, such as financial institutions. The business processes and market actors observed in the PV energy generation sector are discussed in further detail in Appendix C.

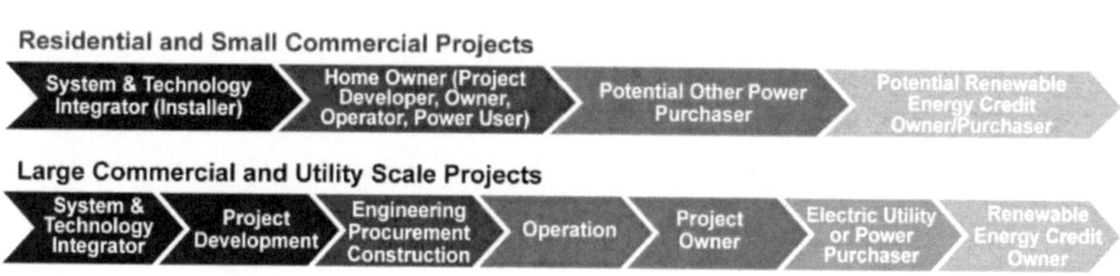

Figure 9. PV energy generation sector value chain.

1.2.1. Project and Service Distinctions

PV energy generation sector services are differentiated most by their scope, scale, the degree to which their services are integrated or bundled together, and the degree of project risk and complexity:

- *Scope.* PV energy generation sector services range from performance modeling, to tax credit investing, to construction management. Although this wide range of services is necessary to execute large projects, many services can be geographically distant from project installation sites.
- *Scale differentiation.* PV energy generation companies typically focus their business models on supporting either residential-scale (typically less than [<] 10 kW), commercial-scale (10 kW–1 MW), or utility-scale (greater than [>]1 MW) projects, with some crossover. Scale differentiation is clearest at the utility scale, where the project size drives a significantly more complex process and more specialized services.
- *Level of integration.* PV energy generation companies often integrate the many required steps of executing a PV project into a turn-key package, taking ownership of a project from start to finish with a single contract. The individual tasks required can all be done in-house, or be distributed in subcontracts. This integration is most common in physical design and construction, especially at large scales, but can include other financial or legal activities.
- *Risk.* Projects on challenging sites, or with complicated use requirements or uncommon ownership arrangements, have increased project risk and typically are handled by companies with relevant specific expertise. Examples of specific expertise include projects on brownfield sites, on tribal lands, or in extremely remote or hazardous conditions.

1.2.2. Costs

Like most service costs, PV energy generation services costs in the United States vary by geography, project scope and scale, the level of expertise or specialization required, and the level of competition. The costs of services are usually built up from hourly rates, and calculated for comparison purposes in dollars per watt at the project level.

PV energy generation sector costs are not driven down directly with economies of scale, but design, permitting, finance, and legal services' share of project costs generally do go down as project size increases.

1.2.3 Trends

1.2.3.1. Increased Third-Party Ownership Accompanying Power Purchase Agreements

Companies with experience in finance (e.g., banks and insurance companies) are increasingly bearing the up-front costs of a PV project in return for the secure future cash flows and tax benefits that can be created through a power purchase agreement (PPA) with the local electric utility or the host of the PV system.

By separating ownership and hosting of the PV system, this PV project structure creates benefits for multiple parties. Third-party financiers take on some project risk, but they monetize the financial incentives being offered, along with the long-term cash generated according to the terms of the PPA. Third-party financiers typically also have more ready access to cheaper sources of capital as well as the ability to monetize tax credits ("tax appetite") than other potential project developers or project owners.

1.2.3.2. Increased Participation of Utilities

The economics of PV in the United States to date have not driven utility ownership and operation of PV energy generation projects. Utilities are currently required to "normalize" the accounting for tax benefits (such as the solar investment tax credit or accelerated depreciation). This means that the value of the tax benefits passed through to utility customers is spread over the life of the system, rather than the time the tax benefits are incurred, making the levelized cost of energy from the system appear higher for utilities than for third-party PPA providers who are not required to normalize the accounting for tax benefits.

PV is also inherently difficult for utilities to operate and absorb because of its inconsistent power generation. For these reasons, many utilities, even those facing renewable portfolio standard (RPS) requirements, are electing to purchase renewable energy certificates (RECs) from project developers or project owners rather than own large PV generation projects themselves. However, as the industry continues to mature and PV product costs and perceived risks continue to decrease, utility PV project ownership, and utility PPA agreements will increase.

1.2.3.3. Continued Integration

The PV energy generation sector is becoming increasingly integrated, with many of the largest system integrators, project developers, and engineering, procurement, and construction (EPC) firms expanding their scopes and roles on large projects. This integration is also crossing over into the PV products sector.

At smaller scales, many energy generation companies are finding integration advantages across geographies. By replicating a successful business model like third-party ownership, many system integrators are working with local contractors to build regional teams to take advantage of local market conditions.

1.2.3.4. Differentiation and Diversification

The PV energy generation sector is maturing. It is slowly defining various market niches based on local economic conditions and solar resources, and shaping services that best fit them. PV energy generation sector companies are increasingly finding advantages in developing local expertise and relationships. The most important local drivers of PV energy generation sector attractiveness are discussed below.

1.2.4. Site Selection Criteria

When PV energy generation sector companies choose a new market to enter, they evaluate areas based on the key criteria listed below. In forming city-level economic development strategies, communities need to consider how they measure up against these metrics, and what they can do to improve their overall score:

- Solar resource
- Local cost of electricity and utility structures and attitudes
- Public/private support and policies (rebates and incentives)
- Workforce and training.

Table 4. Matrix of PV Energy Generation Site-Selection Market Criteria

	Energy Generation	
Industry Criteria	Residential / Commercial	Utility Scale
Solar Resource	2	1
Cost of Electricity, Utility Structure and Attitudes	1	1
Public / Private Support and Policies	1	1
Workforce and Training	3	3

Legend: Critical (1) Somewhat Important (2) Less Important (3)

Table 4 illustrates the relative importance of these site selection criteria to companies in various parts of the PV energy generation sector. In forming city-level economic development strategies, communities can use these criteria to quickly evaluate themselves relative to the needs of the industry segment they are considering targeting. In so doing, they can identify potential strengths, challenges, and criteria on which they must improve to be successful. Note: the rankings in Table 4 are solely the opinion of the authors. These rankings are a first-order attempt at characterizing the siting criteria utilized in the PV energy generation sector and are included here solely for the purposes of economic development planning.

2. Employment Opportunities and Economic Development Strategies

2.1. Employment Opportunities

2.1.1. PV Products Sector

As the PV market grows, most analysts expect the solar supply chain to produce many more jobs than are available today. Optimistic analysts estimate that solar PV installed capacity will increase dramatically in the next 20 years, bringing potentially a tenfold or greater increase in solar-related employment opportunities (Worldwatch Institute 2008, Global Insight, Inc. 2008, and Friedman 2009). However, PV products sector jobs will not scale directly with installed capacity. Imports of PV products will continue, and as the solar PV industry matures, the United States will see increases in manufacturing efficiencies, resulting in fewer jobs per watt (Friedman 2009). See Figure 10. Although the figure does not break out PV products sector jobs specifically, a 2005 report from the Renewable Energy Policy Project suggests that as much as 80% of all PV jobs involve manufacturing and assembly, with the remaining 20% involving construction and installation (Sterzinger and Svrcek 2005).

Figure 11 illustrates baseline forecasts (given little to no added support for the industry) of the number of jobs that could be created if efforts were undertaken to support the PV industry. Because of the different metrics used in the projections, and the different analysis dates, the numbers themselves cannot be compared directly. However, the trend clearly predicts a significant increase in jobs available in the PV industry over the next 20 years.

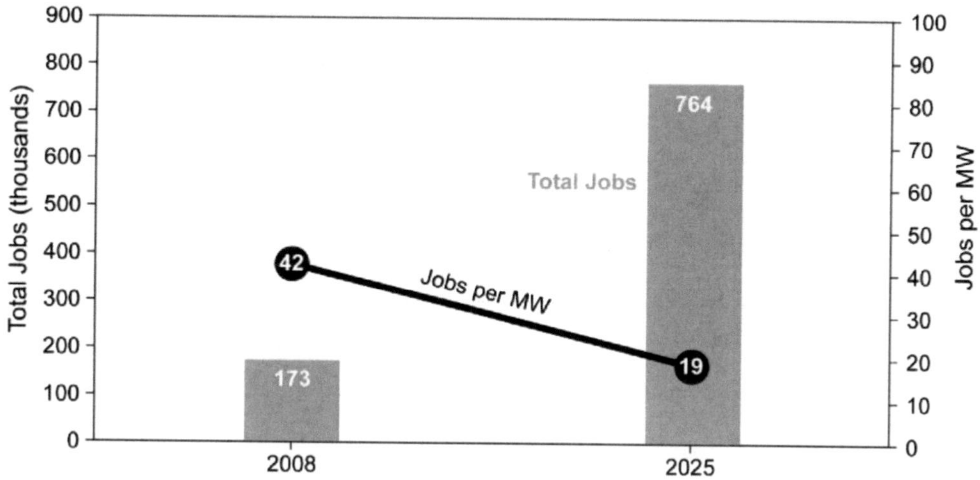

Figure 10. Decrease in jobs per megawatt with increase in total jobs.

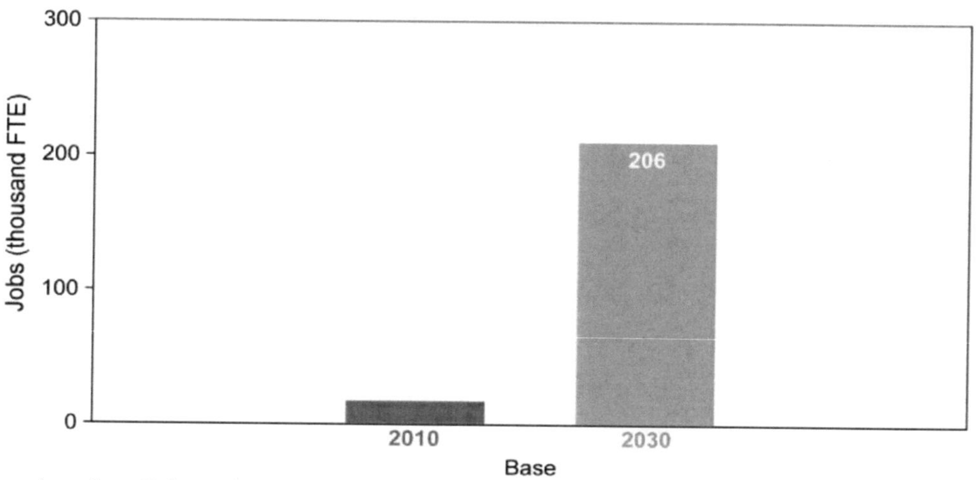

Source: American Solar Energy Society and Management Information Services, Inc. 2008.

Figure 11. High and low forecasts for jobs in the PV industry in the United States.

Tables 5 through 7 estimate the types and number of jobs available in a typical ingot preparation and cell manufacturing facility, module assembly facility, and thin-film production facility. Estimated staffing levels for each type of facility assume production capacity sufficient to generate 65 MW of PV capacity annually.

The PV products sector also creates indirect jobs, such as sales, planning, financing, consulting, and investing. Short-term construction jobs are created for the building of manufacturing facilities. These indirect and short-term jobs are not included in this analysis.

The bulk of PV products sector jobs are in manufacturing operations, specifically in ingot preparation and wafer slicing, and PV cell manufacturing, or in thin-film panel production. These operations require a skilled and trained workforce that includes a large percentage of production operators and assembly workers.

The typical PV products workforce consists predominantly of production line operators, but it also includes line supervisors; industrial, mechanical, and chemical engineers; quality managers; procurement, logistics, and material managers; and other management and administrative positions.

Table 5. 65-MW Manufacturing Facility Staff Mix

Staff Mix	FTE Staff
Production Operators and Maintenance Technicians	120 - 220
Engineers	5 - 10
Management / Administration	5 - 10
Total Workforce	130 - 240

* Labor workforce size and mix is heavily dependent on specific technology and number of process steps. Above figures are general averages only. It should be noted that published sources vary widely between companies and technology.

PV module assembly is much less costly and complex than ingot preparation and slicing or cell manufacturing. A module assembly workforce requires general manufacturing skills, with some special skills dependent on the technology and processes.

Table 6. 65-MW Facility Module Assembly Staff Mix

Staff Mix	FTE Staff
Assemblers	80
Engineers	10
Management / Administration	10
Total Workforce	100

Thin-film PV manufacturing steps cannot be broken into separate factories like ingot preparation and slicing, cell manufacturing, and module assembly, as they represent the equivalent of all of these steps. Thin-film staff counts, while potentially lower per megawatt output, generally have a higher average skill level and wage than c-Si-based processes.

Table 7. 65-MW Facility Sample Thin-Film Staff Mix

Staff Mix	FTE Staff
Production Operators and Maintenance Technicians	80 - 120
Engineers	10 - 15
Management / Administration	5 - 10
Total Workforce	95 - 145

* Labor workforce size and mix is heavily dependent on specific technology and number of process steps. Above figures are general averages only. It should be noted that published sources vary widely between companies and technology.

2.1.2. PV Energy Generation Sector

As mentioned earlier, as few as 20% of all jobs in the PV industry come from the energy generation sector, the majority of which are in installation. This disparity in job opportunities by sector is largely due to the duration of each type of job. The manufacturing and assembly jobs involve long-term operation of facilities, while installation of solar energy systems provides short-term employment for installers. Unless there is a steady stream of new projects, installation does not provide long-term employment. Utility-scale solar power plants employ relatively few long-term workers.

The PV energy generation industry also creates indirect jobs, such as sales, planning, financing, consulting, and investing. Indirect PV generation jobs were not included in this analysis because they are difficult to quantify, largely because the definitions of direct, indirect, and induced jobs vary between studies (Wei, Patadia, and Kammen 2010).

Installation jobs primarily require skills as an electrician, construction laborer, site and civil earthwork laborer, or roofer with some specialized training and certification. Workers currently in the construction industry can supplement existing skills with solar or renewable energy classes or apprenticeships with experienced solar installers (Environmental Defense Fund and Ella Baker Center for Human Rights 2008). Community colleges can support their local workforce by developing programs in solar and renewable energy. Organizations such as the Interstate Renewable Energy Council (IREC) audit and certify these programs, many of which work in conjunction with apprenticeship programs designed specifically for green jobs (State of New Jersey, Department of Labor and Workforce Development 2010, State of California, Department of Industrial Relations 2010, Interstate Renewable Energy Council, 2010).

The location of residential and commercial PV installation jobs is driven by local incentives, solar resources, and the competing price of electricity. Utility-scale installations may be designed by non-local companies, but local installers and construction personnel will more than likely do the physical work. Areas wishing to increase installation job opportunities should focus on improving the local market for the systems by gearing incentive programs to the consumer.

Table 8 provides estimates of full-time equivalent (FTE) staff for installation of a single, 50- MW, utility-scale PV system.

Table 8. Sample 50-MW Utility-Scale Installation Staff Mix

Staff Mix	FTE Staff
Craft Labor	175 - 200
Professional Labor	75 - 100
Total Workforce	250 - 300

2.2. Economic Development Strategies

2.2.1. PV Products Sector

Not all cities are good locations for all segments of the PV products sector. Local economic development groups must consider and analyze local assets and challenges, and target PV industry segments that are attainable. Appendix B lists the typical criteria that PV products manufacturers consider when selecting a manufacturing site.

2.2.1.1. Designing a Successful Economic Development Strategy

Successful economic development programs have been both comprehensive and strategically focused on those segments of the products sector for which the community's assets and infrastructure are a good match. In many cases, successful economic developers were also able to identify their disadvantages and invested to overcome them.

Successful economic development programs targeting the PV products sector industry have several common elements, including:

- Investing in understanding the requirements for the various segments of the industry and the relevant local assets and challenges. Matching an area's attributes to the needs of targeted businesses creates more achievable goals.
- Determining those disadvantages that can be overcome and those that cannot
- Developing incentives specifically to overcome area disadvantages, or to provide a competitive advantage
- Providing non-financial assistance that can improve the business environment
- Taking a long-term approach. Although there is usually pressure to achieve short-term results, successful economic development programs must think long term
- Working with higher education and research to support and complement the targeted industry segment. These efforts should allow for collaboration with industry and be open to partnerships and intellectual property development.
- Marketing efforts to targeted segments only after careful analysis.

2.2.1.2. Economic Development Tools and Programs

States, regions, and cities have developed a wide range of tools, both financial and non-financial, to support and grow the solar PV products sector in their areas.

Financial Tools

A successful, comprehensive economic development program for the PV products sector that typically includes financial incentives.[2]

Grants

Grants are powerful incentives that can be used to assist directly with the development or expansion of a new facility, to fund research projects into new technologies, or to conduct other activities that directly benefit a targeted segment. Grants for new facilities are most often provided on a case-by-case basis, with the amount of the grant based on job generation, investment, and meeting and expanding the economic goals of the area. The granting entity and business typically enter into an agreement that describes terms of the grant and penalties for not meeting the terms.

Grants are often viewed by the PV industry as an effective way to reduce investment requirements. However, grants programs offer little opportunity to sustainably operate long term, as they typically have no pay-back requirements.

Loan Programs

With direct loan programs, money is lent directly from the lending entity to the project owner. Direct loans allow money to be lent at low interest rates to projects that might not receive funding through traditional financing. They can be easier on the lending agency than rebates and grants because payments and interest can be invested back into the fund. Direct loans require higher capital, more risk, and greater administrative costs by the agency than other types of loans.

Matching loans provide a share of a loan, often at low or no interest, to borrowers who are able to find traditional financing from a commercial lender. These loans reduce the risk and expense to the granting agency, but rely on private lenders' willingness to match the loan.

Interest rate buy-down programs and linked deposits both subsidize rates offered by commercial lenders, resulting in a lower interest rate to the project owner. Interest rate buy-down reduces costs for project owners while alleviating the risk of loan defaults for the lending agency, but— like rebates and grants—the lending agency does not receive any repayment. These programs also hinge on the owner's ability to secure financing through a commercial lender.

Loan guarantee programs have received a great deal of attention lately in conjunction with the American Recovery and Reinvestment Act (ARRA). Several large manufacturing projects have received loan guarantees through the U.S. Department of Energy that have allowed them to secure investment funding. Guarantees will continue to be important because they protect lenders against a specified amount of capital losses if a borrower defaults on a loan. Loan guarantees can allow projects that normally would not qualify for financing to receive loans. This is especially useful to manufacturing companies scaling up production to bring new technologies to market.

Many states and cities offer direct loans, reduced-interest loans, and matching loans specifically targeted for manufacturing projects. These loans may also contain clauses that allow them to be forgiven if certain conditions are met. These conditions usually include job creation and investment within an agreed-upon timeframe. These loans often include a cap that requires that additional equity or debt financing must be secured, and can be used for construction financing. These loans can be strong incentives for companies to locate manufacturing facilities in particular areas.

Tax Incentives

PV products sector tax incentives are granted by states and local governments to companies that move to, or expand in, a certain area. These usually take the form of reduction or exemption of sales tax, employment tax, corporate income tax, or property tax. Many times these are granted outright, but they can also be provided as tax credits to reduce tax liability in the short or long term. Additional credits are offered in some areas for R&D expenditures, employment of certain workers, and investment tax credits.

Tax credits that can be monetized or sold to others enable more businesses to take advantage of the credits. This is a particular advantage to companies that may not have a tax liability in a location for several years and cannot effectively use the credit. The ability to sell the credit can provide up-front investment capital, which is especially valuable to startup companies.

Other Financial Incentives

Most states and cities offer customized incentive packages, on a case-by-case basis, that have the potential to provide a significant number of jobs or otherwise contribute strongly to the local economy. Examples of incentives include the following:

- Discounted or free land, infrastructure upgrades, and waived or reduced development fees
- Research funding or collaboration opportunities with research universities or institutes, and workforce training targeted to the specific needs of the company
- Reduced electric or other utility prices or long-term price contracts
- Off-take agreements to guarantee markets and lower risk (an off-take commitment is a commitment by a public or private entity to purchase a portion of the new facility's production over a set period of time)
- Company-specific incentives can have high up-front costs, but economic development groups typically can analyze the payback and the competition by other regions before making a commitment. Specialized incentive packages are generally expected by large PV products sector companies and can be the deciding factor in site selection.

Non-Financial Tools and Programs

Successful economic development programs have incorporated non-financial incentives and programs, such as:

- Improvements to permit processes, reasonable permitting fees and licensing costs, and a supportive regulatory environment
- Overall tax policies supportive of the PV products sector
- Involvement and commitment of public and private sector leaders, including active, visible champions, elected officials, utility and business leaders, public servants, and business and neighborhood organizations
- Public outreach and education on the broad benefits of the PV solar industry
- Targeted marketing to appropriate industry segments.

Technology Development Assistance Programs

PV product companies are driven to constantly improve their product performance and reduce costs through R&D and collaboration.

Areas with technical resources, local research laboratories, research universities, and specialized institutes have used these assets as an incentive to attract and grow PV product companies. Access to existing local expertise is especially important to companies in the early stages of growth. Frequently, the benefits of technology development programs are not one sided, and local institutions greatly appreciate industry connections.

2.2.2. PV Energy Generation Sector

Similar to the PV products sector, not all areas are prime locations for segments of the PV energy generation sector. Local economic development efforts must consider and analyze local assets and challenges, and target PV industry segments that are attainable. Appendix C lists the typical criteria that businesses in the PV energy generation sector consider.

2.2.2.1. Designing a Successful Economic Development Strategy

Areas that have had success in growing jobs in the PV energy generation sector have created comprehensive programs that have a strategic focus based on sound understanding of industry requirements and local conditions. In many cases, these areas have assumed disadvantages where there are none. One clear example of this lack of industry understanding is the common assumption that PV is not well suited to northern latitudes and cold climates.

Programs that have had success in attracting or growing segments of the PV energy generation sector have several common elements, which often include the following:

- Investing in understanding the requirements of various PV energy generation markets and the relevant local assets and challenges. Matching an area's attributes to the needs of targeted businesses allows increased focus on the most realistic opportunities.
- Determining what disadvantages, such as restrictive PPA or net metering rules, can be overcome through public policy change to create an attractive market. Many states do not currently allow PPAs, which is a significant barrier to the development of PV in the United States.
- Developing financial incentives that can be used to overcome area disadvantages like lessthan-ideal solar resources, and that provide a competitive advantage
- Providing non-financial assistance that can improve the business environment and make investment and development easier
- Focusing on a specific PV segment(s) as a strategic target for economic development and creating a supportive public policy environment
- Taking a long-term approach. Although there is usually pressure to achieve short-term results, successful areas have leveraged early actions into long-term results.
- Making investments as needed in workforce training, education, and research that support and complement energy generation sector needs. These efforts should allow for collaboration with industry and be open to partnerships.
- Marketing efforts to targeted segments based on careful analysis.

2.2.2.2. Economic Development Tools and Programs

States, regions, and cities have developed a wide range of financial and non-financial tools to support and grow the PV energy generation sector in their areas. To simplify the discussion of these tools and programs, they are divided below into financial and non-financial tools.

Financial Tools

A successful comprehensive program needs to include tools and financial incentives that provide direct assistance. The following are the most commonly used tools.[3]

Rebates

Rebates are payments made to a PV energy generation project owner, after the project is complete, to cover part of the capital costs. Most often, rebates have specific criteria for size and performance of systems, requirements for certified installers, and sometimes requirements for local content, other energy-efficiency measures, or other restrictions. The amount of the rebate is usually based on system size, but it can also be a fixed dollar amount or percentage of total dollar amount per system. Rebates can be structured to draw from a limited funding amount, or confined to a period of time, and then renewed or adjusted based on changes in the market, new policies, or program goals.

If not well structured, rebates can create market dependency that leads to a dramatic drop in demand when rebate programs end or rebate amounts are reduced. They also are not revolving or self sustaining, and once rebates are awarded to the customer, there is typically no return back into the fund.

Production-Based Incentives

Production-based incentives (PBI) are tied to actual or forecasted energy generation in kilowatt-hours for a fixed period of time. They can take the form of rebates or installment payments over a year or more. PBIs may be based on the actual performance as measured by a monitoring system, or on estimated output based on size, efficiency rating, and site as determined by a credible system performance model. Unlike rebates, PBIs grant the greatest incentives to the systems that perform the best (not simply to the largest nameplate sizes), encouraging careful planning and quality installations. While PBIs may make more sense for the funding agency, they can be less valuable to project owners at small scales. Although they do accelerate return on investment, they do not help with up-front project costs. They also require ongoing tracking and payment, which increase the administrative costs of the program.

Incentive or tariff rates can be set based on type of technology, project size, location, type of ownership, or any other variable that the agency is promoting.

These types of incentives require close coordination with utilities, as prices are determined and the programs are fine tuned and adjusted over time to match the market. Because it is difficult to predict the response to new PBI or feed-in tariff programs, determining the best rate can be challenging. Too high a rate could lead to rate increases if the number of generation projects surged, but a rate that is too low could fail to yield enough new projects. Feed-in tariffs also require that net metering and interconnection policies be in place and well structured.

Grants

Grants for energy generation sector markets usually target installation projects at a residential or commercial/utility scale, within a specific boundary, or on a property with a specific prior or ongoing use. Large grants are usually awarded to qualified projects on a competitive or request for proposals (RFP) basis. Grants are an effective way to reduce investment requirements. However, they offer little opportunity to operate sustainably long term, as they typically have no pay-back requirements.

Loan Programs

There are also a wide variety of loan programs: direct loans, matching loans, interest rate buy-downs, linked deposits, or clean energy assessment districts.

With direct loan programs, money is lent directly from the lending entity to the project owner. Direct loans allow money to be lent at low interest rates to projects that might not receive funding through traditional financing. They can be easier on the lending agency than rebates and grants because payments and interest can be invested back into the fund. Direct loans require higher capital, more risk, and greater administrative costs by the agency than other types of loans.

Matching loans provide a share of a loan, often at low or no interest, to borrowers who are able to find traditional financing from a commercial lender. These loans reduce the risk and expense to the granting agency, but they rely on private lenders' willingness to match the loan.

Interest rate buy-down programs, loan loss reserves created with public funds and linked deposits can result in subsidized rates offered by commercial lenders, resulting in a lower interest rate to the project owner. Interest rate buy-down reduces costs for project owners while alleviating the risk of loan defaults for the lending agency, but—like rebates and grants—the lending agency does not receive any repayment. These programs also depend on the owner's ability to secure financing through a commercial lender.

Clean energy assessment districts and property tax financing, also called Property-Assessed Clean Energy (PACE) models, all eliminate the up-front cost of installing a system by creating an installment payment plan tied to utility payments or property taxes. These programs also reduce the risk to owners by tying the payments to the property. For example, when the property is sold, the new owner takes over the payments along with the utilities and property taxes. These programs can be administratively complex. As of August 2010, the residential PACE model has been put on hold, as rulings from federal financial regulators have specifically opposed any programs where special property tax assessments are in a senior lien position. While the PACE model is very promising, the future of PACE models as a financial tool is unsure.

Tax Incentives

Energy generation incentives are often in the form of tax credits, where a certain percentage of installed cost or dollar amount can be regained as a state or federal tax credit or deduction. States with sales tax often offer sales tax exemptions on the purchase of PV systems, effectively giving the consumer a discount on the system price. Some tax incentives also extend to property tax exemptions for residential and commercial owners. Tax credits and sales tax exemptions can apply to small or large systems, while property tax exemptions generally do not apply to utility-scale projects.

Tax credits that can be monetized or sold to others enable more businesses to take advantage of the credits. This is a particular advantage at the utility scale and for projects with specific tax investors or developers that do not have a tax liability in a location for several years.

Non-Financial Tools and Programs

Although specific financial incentives are very important tools, they are not the only options available. Successful areas have integrated economic development tools and incentives into a comprehensive program of support. Program elements that successful areas have used include the following:

- Policies and a regulatory environment that support market transformation that can include easy and understandable permit processes, expedited processes for interconnections and incentive applications, and reasonable permit fees

- Overall tax policies that support PV installation
- Involvement and commitment of public and private sector leaders in the promotion and support of energy generation projects. These include elected officials, utility and business leaders, public servants, and business and neighborhood organizations
- Public outreach and education on the broad benefits of local PV energy generation can be one of the most important factors in gaining public support for efforts and incentives
- Targeted marketing to appropriate installation types and applications.

Available economic development tools that are specific to the PV energy generation sector include RPSs, and allowances or requirements for net metering, PPAs, and third-party leases.

Renewable Portfolio Standards, Renewable Energy Certificates, and Solar Set-Asides

RPSs are requirements by the state or local jurisdiction that local utilities supply a certain percentage of their energy from renewable resources. These programs may treat all renewable energy sources as equal, or provide different levels of support for different energy sources, called "set-asides." Set-asides (also known as carve-outs) require that a portion of the RPS must use specific technologies or project types. Utilities may comply with the RPS by direct investment in renewable generation projects such as solar PV, by purchasing solar RECs, or by making compliance payments to the requiring agency.

RPS solar set-asides are one of the most powerful and influential drivers for utilities to increase use of PV power. These set-asides are also often drivers of other supportive solar policies discussed in this section.

Net Metering and Interconnection Standards

Policies such as interconnection standards and net metering rules allow system owners to connect to the grid. They govern how owners will be compensated for energy sent to the grid over and above what they have consumed.

Net metering and interconnection rules are often driven by state governments, which generally authorize their state public utilities commissions to develop net metering rules. At the city level, city governments can influence municipal utilities over net metering rules in their territories.

Net metering is one of the most important policy drivers for PV markets. This is because it enables system owners to reduce energy costs, and it is an enabler for other financial incentives (such as feed-in tariffs).

Third-Party Lease Agreements and Residential Power Purchase Agreements

Third-party leases and residential PPAs typically are provided by private sector companies. However, support for these companies can be a positive influence for residential PV market transformation by allowing residential and commercial property owners to have systems installed with reduced up-front cost and maintenance requirements. The simplest arrangement is typically structured in the following way:

- PPAs
    - The building owner agrees to host the system on its roof and purchase the electricity generated by the system over an extended period of time.
    - The system owner agrees to lease the roof space and to install, operate, and maintain the system.
    - The solar developer/PPA provider is usually partnered with a financial investor who takes the tax benefits associated with the PV installation.

- Solar Leases
    - Under a solar lease, the building owner leases the system from the solar developer. The monthly lease payments that the building owner makes are offset by the utility bill savings.
    - At the end of the lease, the building owner often has an option to purchase the system, extend the lease, or have the system removed.
    - The solar leasing company is usually partnered with a financial investor who takes the tax benefits associated with the PV installation.

One potential complication in these types of arrangements is that the system is generally attached to the property, but the lease or PPA is attached to the owner. This puts the burden on the property owner to transfer the lease if the property is sold, which may be difficult.

Community PV

Community-based PV programs organize and enlist participation from residential and small commercial owners to collectively install a large PV system in coordination with a local utility, or to install distributed smaller systems under a single contract.

In the case of a centralized system, participants buy into the program and benefit from the system-generated energy in proportion to the amount of their buy-in. In the distributed model, each participant purchases an independent system as part of a group buy. The key advantage to these models is that they can reduce system costs through larger scale purchases and allow non-homeowners to contribute to and benefit from a PV system.

These programs typically are initiated, organized, and managed by community groups, local utilities, or local businesses. If not organized by the utilities themselves, centralized community solar projects require close collaboration with utilities and with administrators of net metering policies, as well as careful legal crafting to avoid the formation of investments that might trigger state or federal securities law. Public support for these projects is important because it can increase the number of installations and related job growth.

3. Case Studies and Examples

Many communities have been successful in developing various segments of the PV industry. The following are examples of incentive programs and public policy commitments by different jurisdictions, and their results.

3.1. PV Products Sector

When economic developers target the PV products sector, what they are typically targeting is a large factory and the factory's centralized, stable, direct, and indirect jobs. Because these opportunities are generally large, it is often beyond the resources of cities to offer substantial enough incentives that are competitive. As a result, city economic developers lean heavily on state governments for large opportunities. The following examples contain mostly state incentive and program details; however, city-level efforts are noted whenever applicable.

3.1.1. New York

3.1.1.1. Program Features

- The Solar Energy Consortium is a non-profit public/private partnership focused on attracting and growing solar companies, and increasing PV efficiency through collaborative industry/university R&D initiatives.
- New York State Energy Research and Development Authority (NYSERDA) provides funding to New York-based renewable energy manufacturing companies. The Clean Energy Business Acceleration program provides:
  - Incentives to help fund new production facilities *for clean energy technologies and* commitments of $35 million to support business expansion, including marketing, business development, capital raising, and building business infrastructure
  - Investment of $9 million in the creation of six incubators throughout the state providing mentoring, technical and business assistance, networking, training, access to capital, and other resources to entrepreneurs and start-up clean energy companies
  - Entrepreneurial training for university researchers, mid-career executives, graduate students, and veteran entrepreneurs.
- Empire State Development Corporation provides loans, grants, and tax credits to companies that will locate operations in-state. Incentives are based on private investments, business viability, and job creation potential.
- Supportive public policies include an RPS of 30% by 2015, net metering, and commitments to invest $2.99 billion into renewables projects through 2025.
- Focus on education and training for the solar industry is provided through a consortium of New York community colleges and the New York Department of Labor.
- The state is focused on developing an integrated in-state PV manufacturing supply chain.

3.1.1.2. Successes

New York's comprehensive economic development approach has resulted in nearly 1,000 direct new manufacturing jobs and dozens of companies in the state solar supply chain.

Empire State DevelopmentCorporation *Website http://www. .empire.state.ny.us/* The Solar Energy Consortium *Website http://thesolarec.org* New York State Energy Research andDevelopment Authority *Website* www.nyserda.org

- Globe Specialty Metals is converting an idle factory in Niagara Falls into a solar-grade silicon manufacturing facility, creating up to 500 new jobs. The New York Power Authority has provided low-cost hydropower to the company in return for the right of first refusal to 20% of the silicon manufacturer's annual production, which can be offered at a reduced cost to downstream PV companies located in-state. This is the first silicon supply incentive offered by a U.S. jurisdiction.
- Solar Tech Renewables plans to build a PV module assembly facility that will create 300 jobs. It will purchase solar cells produced by New York-based SpectraWatt for use in their modules.
- Prism Solar Technologies has repurposed a former plasma display factory for the manufacture of unique solar concentrator films and modules. The company expects to employ up to 250 people in the next 5 years and has hired many of the laid-off workers from the plasma display facility.

3.1.2. Colorado and the City and County of Denver

3.1.2.1. Program Features

- Incentives and assistance programs offered to manufacturing companies including tax credits, grants, and loans from a strategic fund, and job training and investment tax credits available to small innovative companies
- State and local focus and public policy support for solar and renewable energy projects, increasing the local market for PV products; these include a RPS that was recently expanded to 30% by 2020, net metering policies, and financial assistance for installations
- Proximity to R&D and higher education from the National Renewable Energy Laboratory (NREL) and numerous Colorado Universities that have formed the Colorado Renewable Energy Collaboratory to support and promote renewable energy research
- Existing technology companies in Colorado have provided a base of skilled workers and a technology-oriented manufacturing environment
- Technical job training programs
- The *Metro Denver Economic Development Corporation (Metro Denver EDC)* works with the state and its local partners to help PV products sector companies select sites, expedite permitting, and negotiate state, city, and county incentives and job training assistance programs.

Colorado Office of Economic Development and International Trade *Phone*: (303) 892-3840 *Website:* http://www.colorado.gov/cs/Satellite/0EDIT/OEDIT/1162927366289 Governor's Energy Office *Phone*: 28003 632-6662 *Email*: geo@state *Website*: http://www.colorado.gov/energy Metro Denver Economic Development Corporation *Contact:* Laura Brandt *Phone*: 303.620.8092 *Email*: laura.brandt@metr*o*denver.org *Website*: http://www.metr denver. org.

3.1.2.2. Successes

- PrimeStar Solar, a manufacturer of thin-film PV modules for large-scale commercial applications, has expanded to a 106,000-square-foot facility to increase production. The Metro Denver EDC assisted with site selection, while the City of Arvada and Jefferson County provided incentives in the form of business personal property tax rebates.
- SMA Solar, a large manufacturer of inverters for solar power systems, has developed a manufacturing facility in Denver that will initially employ 300, with potential to grow to 700. The state offered incentives in the form of tax credits, job training assistance, performance incentives, and enterprise-zone tax credits. The City of Denver has also offered job training, relocation expenses, and site selection services.
- *Abound Solar*, a manufacturer of cadmium telluride (CdTe) thin-film PV modules, received a $400 million loan guarantee from the DOE in 2010 to expand its manufacturing facilities in the Denver metro area.
- *Ascent Solar*, a manufacturer of thin-film PV modules, opened its new world headquarters in March 2009 in Denver. Ascent will employ 300 people at the 145,000-square-foot plant, with plans to add an additional 200 employees in 2 years.
- *The Solar Technology Accelerator Center* is a facility based in suburban Aurora, Colorado, for the testing of solar products and new innovations. The city of Aurora assisted with the acquisition of land and infrastructure for the center.

3.1.3. New Mexico

3.1.3.1. Program Features

- Green job training programs at state community colleges and incentivized job training programs customized for qualifying companies
- Creation of a governor-backed Green Jobs Cabinet, tasked with developing a statewide strategic plan for clean technology economic development

- The state's *Job Training Incentive Program* reimburses qualified companies for on-the-job and classroom training for creating qualifying jobs. The reimbursements range from 30% to 75% of the employee's wage for up to 1,040 hours. Additionally, specialized classroom training at post-secondary educational institutions is eligible for $1,000 in reimbursement per employee hired through the program
- Numerous highly leveraged tax credits intended to stimulate both the PV products and PV energy generation sectors, such as the *High Wage Jobs Tax Credit, the Alternative Energy Product Manufacturers Tax Credit, and the Advanced Energy Tax Credits, the Renewable Energy Production Tax Credit, as well as Industrial Revenue Bonds* to finance new machinery and equipment
- *R&D centers and educational research programs* at Sandia National Laboratories and research universities, such as University of New Mexico, New Mexico Tech, and New Mexico State University all have expertise in technologies at the core of PV products. Close proximity to these permits PV products sector companies to capitalize on breakthrough research while allowing researchers to benefit from industry collaboration
- The *New Mexico Renewable Energy Transmission Authority* is authorized to develop new transmission lines to serve the renewable energy installations in New Mexico, increasing the potential for siting of large-scale solar generating facilities.

3.1.3.2. Successes

- SCHOTTSolar has completed a manufacturing facility in Albuquerque to produce PV modules and receivers for concentrating PV solar power plants. The 200,000-square-foot facility employs 325 workers.
- In a joint venture between Germany-based Fraunhofer, Canada-based CSA Group, and Germany's VDE Testing and Certification Institute, the CFV Solar Test Laboratory will provide testing services for the certification of solar products in North America and internationally.
- Established New Mexico solar companies, including Emcore, UniRac, Array Technologies, Qnuru, and Zomeworks, continue to expand in-state operations.

3.1.4. Michigan

3.1.4.1. Program Features

- Highly leveraged pre-existing economic assests, such as a skilled manufacturing workforce and existing supply chain and manufacturing infrastructure from the automotive industry
- Investment and commitment to R&D through university and research centers, with a focus on renewables and solar technologies, building on Michigan's significant R&D capacity
- Focus on education and training for the solar industry through the development of the University Research Corridor and investment in job training and retraining programs

- Targeted state and local financial incentives for renewables and solar manufacturing, including:
  - The *NextEnergy* program that provides tax credits and general property tax exemptions for renewables research and manufacturing activities
  - The $109 million *21st Century Investment Fund*, created to leverage private sector investments in new and emerging technology companies
  - Creation of a *Venture Capital and Angels Fund and Angel Investment Incentive* tax deduction to help stimulate more early stage investment in technology-based companies
  - State tax credits for "anchor" technology companies that assist in attracting a supply chain facility within 10 miles of the "anchor" facility
  - The designation of *Renewable Energy Renaissance Zones.* Businesses located within these zones receive exemptions from the state business tax, education tax, personal and real property taxes, and local income tax, where applicable. Tax abatements are available for up to 15 years, with a 25% incremental phase-out over the last 3 years.
  - A solar PV-specific tax credit for manufacturing activities or development of PV energy, systems, or technology. The credit is equal to 25% of capital investments in a new facility in a given year, up to $15 million.

3.1.4.2. Successes

- *GlobalWatt.* The state of Michigan approved a performance-based tax credit valued at $14 million over 7 years to encourage GlobalWatt to expand in Michigan. GlobalWatt plans to invest $177 million for c-Si module assembly in Saginaw, creating up to 500 jobs.
- *Hemlock Semiconductor Corporation* plans to invest $1 billion in a new polysilicon manufacturing facility, creating 190 direct jobs. The project is eligible for the Anchor District incentive, $180 million in local incentives, a federal tax credit valued at $141 million, a Community Development Block Grant valued at $6.9 million, and job-training grants valued at $390,000.
- *Clairvoyant Energy Solar Panel Manufacturing, Inc.* plans to invest $856 million at the former Ford Wixom Assembly Plant to manufacture thin-film PV modules. The project is expected to create over 750 direct jobs over the next 5 years and 4,600 indirect jobs.
- *United Solar Ovonic (UNI-SOLAR)* thin-film PV manufacturer has developed four manufacturing facilities in Michigan and is considering a fifth.
- *Suniva* has received a U.S. Department of Energy loan guarantee to construct a solar cell plant, in addition to approval for a PV Michigan Business Tax Credit of $25 million. Suniva's capacity expansion will create up to 500 direct jobs over the next 5 years.

Michigan Economic Development Corporation
Email: MEDCservices@michigan.org
Website: http://www.michiganadvantage.org/

3.1.5. Oregon

3.1.5.1. Program Features

- Advantageous pre-existing economic assets, including low electricity rates, an attractive corporate tax structure, and proximity to a semiconductor industry providing a trained workforce, a valuable supply chain, and existing manufacturing infrastructure
- A supportive policy environment, including an RPS of 25% by 2025, as well as local programs, such as feed-in tariffs and community solar programs, that incentivize the installation of solar PV systems
- A generous *Business Energy Tax Credit (BETC)* equal to 50% of the incurred capital investment costs for eligible renewable energy manufacturing activities, up to $40 million for each phase of development. The credit features a transfer provision, allowing financing partners to monetize the credit to produce upfront investment capital. Projects are subject to a $300 million cap.
- A package-based economic development approach that pairs the BETC, property tax abatements, certified "shovel-ready" sites, and other financial incentives to drive rapid project development.

Business Oregon
Website: http://www.oregon4biz.com
Oregon Department of Energy
Email: Matt.Hale@state.or.us
Websites:
http://www.oregon.gov/ENERGY http://www. oregon.gov/DOR/

3.1.5.2. Successes

- *SANYO Solar* opened an $80 million, 130,000-square-foot silicon ingot manufacturing facility in Salem, Oregon. The facility created an estimated 150 new jobs, with an average salary and benefit package of $50,000 per employee and an estimated annual payroll of $8.6 million. The new manufacturing facility is located within the Salem Renewable Energy and Technology Park, an economic development zone designated by the city of Salem for renewable energy companies. Thanks to state pre-certification, SANYO was able to bypass any permitting and regulatory siting requirements, and also received tax abatements on construction and equipment.
- *Solar World,* a German manufacturer of solar wafers, cells, and PV modules, acquired a former semiconductor manufacturing facility in Hillsboro in 2009, where

it employs 500 workers, with an estimated workforce of 1,000 employees and a 500-MW production capacity by 2011.

- *Solaicx,* a manufacturer of monocrystalline silicon ingots and wafers in Portland, Oregon, employs 80 workers.
- *Solexant,* a thin-film PV manufacturer, has announced plans to construct a new facility in Gresham, Oregon, employing 170 workers and eventually growing to 1,000.
- *PV Powered,* a manufacturer of inverters, employs nearly 100 workers.

3.1.6. Tennessee

3.1.6.1. Program Features

- A comprehensive Volunteer State Solar Initiative, funded in part through $62.5 million in ARRA funding, to increase the number of in-state PV installations and to bolster PV research and development efforts
- The Tennessee Solar Institute, created under the Volunteer State Solar Initiative, leverages the resources of the University of Tennessee and Oak Ridge National Laboratory to coordinate solar PV research and development and to administer two grant programs through its Solar Opportunity Fund
- Relatively low-cost electricity, economic development resources, and incentive programs available through the Tennessee Valley Authority (TVA)
  - The TVA's Generation Partners Program creates a 10-year production-based incentive for residential and commercial installations of qualified renewable generation systems less than 200 kW. Its Midsized Renewable Standard Offer program provides standard offtake rates for systems between 200 kW and 20 MW.

An aggressive financial incentives strategy that includes:

- A *Jobs Tax Super Credit* that offsets 100% of franchise and excise (F&E) tax liability for up to 20 years for capital-intensive projects investing over $100 million and creating at least 100 jobs
- An *Integrated Supplier and Customer Tax Credit* that provides Super Credit tax status to suppliers and customers that locate near a large "anchor" project
- A *Green Energy Tax Credit* for certified green energy supply chain manufacturers, which provides a F&E tax credit for the differential cost to purchase green energy o A *Carbon Charge Tax Credit* allows for a refundable credit for any future carbon charge, whether federal or state, imposed on a certified company o An *Emerging Industry Tax Credit* and a reduced electricity tax rate of 1.5%.

State of Tennessee Energy Policy Office
Phone: (800) 342-1340 or (615) 741-2994
Tennessee Valley Author ity Economic Development
Phone: (615) 232-6225 *Email*: econdev@tva.com
Tennessee Department of Economic and Community Development
Energy Division
Phone: (615) 741-2994 *Fax*: (615) 741-5070
Tennessee Valley Authority
Phone: (865) 632-2101
Email: tvainfo@tva.com
Website: http://www.energvright.com/
Tennessee Solar Institute
Phone: (866) SOL-AR90
Email: s olar@tennessee.edu
Website: http://solar.tennessee.edu

3.1.6.2. Successes

- *Hemlock Semiconductor Corporation* is constructing a large polysilicon manufacturing facility that will create up to 500 new jobs.
- *Wacker Chemie,* a global German manufacturer of polysilicon, has announced the development of a $1 billion polysilicon plant that will create 500 jobs.
- *Confluence Solar* has announced plans to build a $200 million single crystal ingot manufacturing facility for solar cells that will create up to 250 jobs.
- *Sharp Solar Energy Solutions* has continued to expand its PV manufacturing facility and currently employs over 500.

3.2. PV Energy Generation Sector

When economic developers target the PV energy generation sector, what they are typically targeting are installations and the associated energy, RECs, and, most importantly, direct construction jobs required to install PV systems. The larger and more significant programs often require the financial and public policy support of county and state governments. Following are both city- and state-level economic development program examples from throughout the United States.

3.2.1. New Jersey

New Jersey is second only to California in installed capacity in the United States. With a lower than average amount of solar radiation, New Jersey used policy to encourage installation and drive job growth in the PV energy generation sector.

3.2.1.1. Program Features

- New Jersey's Clean Energy Program, administered through the Office of Clean Energy, is a legislatively mandated initiative of the New Jersey Board of Public Utilities, which provides education, information, and financial incentives for renewable energy systems and energy efficiency measures.
- New Jersey's RPS requires that every utility that provides electricity to residents of New Jersey provide 22.5% of total electricity from renewable sources by 2021.
- The Edison Innovation Clean Energy Manufacturing Fund (CEMF) provides assistance to for-profit entities that manufacture energy efficient and renewable energy products by providing grants and loans. CEMF helps companies become competitive with traditional sources of electric generation.
- Interconnection and net metering standards make it easier for systems to connect to the distribution system and to be compensated for their contribution.
- Solar Loan Program allows residents to take a loan from the utility providers for up to 40% to 60% of capital cost of the PV system. Customers may then repay the loan using solar RECs produced by their system.

3.2.1.2. Solar Renewable Energy Certificates

The solar REC program is a solar credit trading system and allows electric suppliers to buy cates in order to meet its solar RPS requirements. The SREC price averages approximately $100 lower per megawatt-hour than the Solar Alternative Compliance Payment during a given year; however, actual prices vary with market demand.

Public Information - Renewable EnergyProgram, New Jersey Board of Public Utilities, Office of Clean Energy
Phone: (866) 657-6278 *Website*:
http://www.nicleanenergy.com/renewable-energy/home/home
Ediison Innovation Clean EnergyManufacturing Fund
New Jersey Economic DevelopmentAuthority
Phone: (866) 534-7789
Email: EdisonCEMF@nieda.com
Website: http://www.nieda.com/

3.2.1.3. Renewable Energy Manufacturing Incentive

The Renewable Energy Manufacturing Incentive program provides rebates for homeowners and businesses who install solar PV system components such as modules, inverters, and racking systems manufactured in New Jersey from certified local manufacturers. Equipment must be made by program-certified manufacturers, which demonstrate at least 50% of the product cost is associated with facilities located in New Jersey.

3.2.2. Ontario, Canada

Ontario, Canada, passed its Green Energy Act in May 2009 with the intent to encourage new investment in renewable energy creating green jobs and economic growth. Goals included creating 50,000 new manufacturing jobs in the first 3 years of the program. The success of this program in creating manufacturing jobs is great evidence of a strong link between incentives on PV energy generation and installations with local content provisions, and PV products manufacturing growth.

3.2.2.1. Program Features

- One of the most aggressive feed-in-tariffs (FIT) in North America
- A domestic content requirement that is tied directly to the FIT which requires projects 10-kW and larger to have 50% domestic product content and projects under 10 kW to have 40%. Both will increase to 60% in 2011
- Renewable Energy Facilitation Office helps companies, developers, communities, and municipalities navigate through the regulatory approvals necessary for starting and completing projects.

3.2.2.2. Successful Project

Samsung C&T Corporation and Korea Electric Power Corporation agreed to develop and operate solar and wind power projects throughout the province. The plan includes up to 2,000 MW of wind and 500 MW of solar generation capacity. The government attracted the agreement by providing incentives for green energy production. Samsung has agreed to build four local manufacturing plants to meet the requirement for locally-made equipment. The agreement is expected to create 16,000 jobs (direct and indirect) from 2010 to 2015.

3.2.2.3. Status Update

- Due to the very large demand for ground-mounted systems less than 10 kW, the FIT was reduced to $.588 per kilowatt-hour in August 2010. The FIT for roof-mounted systems remains at $.80 per kilowatt-hour.
- Existing Ontario manufacturing companies have expressed concern that use of more expensive green power will increase electricity prices and put the company at a competitive disadvantage, restricting its ability to grow and create jobs in Ontario.
- With the rapid growth in small-scale installations, there is a shortage of trained installers. Training institutions are struggling to grow fast enough to meet demand.

Ministry of Energy and Infrastructure
Email: write2us@energy.gov.on.ca
Website:
http://www.mei.gov.on.ca/en/energy/renewable/?page=osthi
FIT Program/microFIT Program, Ontario Power Authority
Email: FIT@powerauthority.on.ca
Website: http://fit.powerauthority.on.ca/
Ontario Ministry of Economic Development and Trade
Email: info@edt.gov.on.ca
Website:
http://www. ontariocanada.com/ontcan/1medt /en/home en.jsp

3.2.3 Portland, Oregon

Portland has capitalized on strong state-level incentives by providing local education programs, supporting community-led volume purchasing campaigns, and lowering municipal regulatory barriers to solar installations.

3.2.3.1. Program Features

- Incentives and tax credits from the State of Oregon and Energy Trust of Oregon, which dramatically reduce up-front costs
- Extensive training and educational opportunities, including workshops, tours, and a solar expo
- Strong community-based social marketing effort through Solarize Portland, supported by the local government
- Streamlined electronic permit submittal process.

3.2.3.2. Solarize Portland

Solarize Portland is a community-based, volunteer-run effort, made possible with support from the City of Portland's Bureau of Planning and Sustainability and the Energy Trust of Oregon. The program connects neighbors who want to invest in solar energy systems, provides workshops, and arranges for an approved contractor. Neighbors work together to understand options, and receive volume pricing on the panels and installation.

3.2.3.3. Oregon Pilot Solar Volumetric Incentive Rates and Payments Program

Oregon's Volumetric Incentive program pays system owners for the kilowatt-hours generated over 15 years, at a rate set at the time of enrollment. The pilot program has a cap of 25 MW, spread over 4 years. The first two rounds of enrollment were fully subscribed within minutes of opening.

> City of Portland, Bureau of Planning andSusta inab il ity
> *Phone*: (503) 823-7700
> *Website*:http://www.portlandonline.com/bps/index.cfm?c=28534
> Solarize Portland
> *Website*: htto://www.solarizeportland. org

3.2.4. San Francisco, California

San Francisco's Office of Economic and Workforce Development (OEWD) emphasizes vocational skills training with the city's Green Skills Academy, which targets individual low-income neighborhoods.

3.2.4.1. Program Features

The training program recruits participants from four of the city's highest-poverty areas.

- The program promotes attainment of job skills to help participants find employment in green industry.

- Participants in San Francisco's GoSolarSF incentive program for installing a PV system must use contractors who employ staff who have completed OEWD training programs.

3.2.4.2 Group Purchase Programs

San Francisco has also had success with group purchase programs. The city's Department of Environment has worked with several local communities to facilitate group purchases. The city has also worked with One Block Off the Grid to organize residential group purchases.

San Francisco Office of Economic and Workforce Development Website: http://www. oewd. org/ Tra inGreenSF Phone: (415) 575- 4556 Email: traingreensf@sfgoodwill. org

3.2.5 Gainesville, Florida

Gainesville Regional Utilities' (GRU) feed-in tariff program is modeled after Germany's FIT. Space in the program was initially acquired by businesses with plans for large systems, but the program has since been modified to dedicate more kilowatts to smaller residential systems.

3.2.5.1. Program Features

- Residential and commercial systems are eligible for FIT.
- The FIT tariff rate is designed to allow an investor to earn roughly a 5% internal rate of return after taxes and realizing all federal tax benefits.
- GRU purchases power from PV system owners at a set rate over a 20-year contract.
- GRU owns any renewable energy certificates from the energy produced.
- Program limit of 4 MW per calendar year is broken into three classes depending on system size and mounting (roof or ground). For small residential systems, 200 kilowatts are reserved, and 1 MW is reserved for ground-mounted systems.

3.2.5.2. Other Incentives

GRU also offers net metering for PV systems and PV system rebates up to $7,500 based on system size and available sunlight, as well as a solar water heating rebate. Systems that have received these other incentives are not elegible for the FIT.

Gainesville Regional Utilities Website: https://www.gru.com/OurCommunity/Environment/GreenEnergy

3.2.6. Salt Lake City, Utah

In 2009 and 2010, Utah went from double "Fs" to double "As" for Intereconnection and Net Metering in the annual *Freeing the Grid* report (Network for New Energy Choices, Vote Solar Initiative, and Interstate Renewable Energy Council 2010). As a direct result of the involvement and educational efforts of the Solar Salt Lake Partnership, Utah was one of only two states to earn two As. Solar Salt Lake partners (in collaboration with NREL and IREC) engaged and educated key decision-makers about best practices for net metering and interconnection and the value of removing barriers to solar energy. In 2008, the state's legislature revised its net metering policy and in 2009 the Utah Public Service Commission adopted favorable rules for net metering customers. The commission finalized updated Interconnection Standards in 2010, following a 2.5-year docket proceeding.

In 2010, Utah adopted a bill to enable third-party financing for certain eligible entities. As a result of these successes, Salt Lake County is slated to install one of the largest roof-top solar projects in the nation, taking advantage of third-party financing and improved net metering.

Utah Clean Energy
Website: http://utahcleanenergy.org/
Rocky Mountain Power Net Metering
Website: http://www.rockymountainpower.net/env/nmcg.html
Solar Salt Lake Project
Website: http://utahcleanenergy.org/our work/solarsalt lake project
SLC Solar Map
http://www.slcgovsalar.com/

3.2.6.1. Program Features

- Net metering is available to residential systems up to 25 kW and nonresidential systems up to 2 MW.
- Residential and small commercial net metering customers receive kilowatt-hour-for-kilowatthour credits for excess generation; the credits roll over month-to-month during the annualized billing period. Large commercial customers are given an option for the value of their excess generation, to better account for demand charges.
- PV system owners retain the renewable energy certificates associated with the power they produce.
- Removal of barriers to third-party financing created opportunity for a 2.6-MW solar PV system on Salt Lake County's Salt Palace Convention Center.
- An implementation plan for Salt Lake City and Salt Lake County establishes a long-term commitment to solar deployment and facilitates solar adoption in all sectors.
- Education and training is provided for building community, utilities, code officials, real estate and appraisal community, city zoning, historic district commissions, and the general public, other local governments, non-profits, churches, schools, and businesses.

3.2.6.2 Solar Salt Lake Project

The Solar Salt Lake Project, managed by Utah Clean Energy in collaboration with Salt Lake City and Salt Lake County, and funded by DOE's Solar America Communities, plans to facilitate at least 10 MW or 10,000 new PV systems by 2015. The Solar Salt Lake implementation plan outlines core strategies to achieve this goal.

APPENDIX A: RECOMMENDED RESOURCES

The following references and resources provide further information on topics addressed in this report that may be relevant to communities' economic development efforts.

Economic Development Tools

- The U.S. Department of Energy's Solar Powering Your Community: A Guide for Local Governments (2009) is a great resource, covering best practices and lessons learned from the original 25 Solar America Communities. These examples are available as an easy-to-use online resource titled Solar Powering Your Community: A Guide for Local Governments. The guide describes many of the country's most innovative solar programs and policies. It also provides implementation tips and example case studies that are focused mostly on growing PV markets.
- GTM Research published a report titled PV Manufacturing in the United States: Market Outlook, Incentives and Supply Chain Opportunities (Mehta 2009). The report offers an overview and forecast of the PV manufacturing industry, as well as in-depth analysis of the location of PV manufacturing facilities and the types, availability, and success of manufacturing incentives.
- Clean Energy Group has published two guides to assist with local renewable energy efforts. Both are aimed at states rather than cities, but the material described will be helpful to cities as well. Both focus more heavily on the PV energy generation sector, but they will be useful in regard to both sectors.
    - Smart Solar Marketing Strategies: Clean Energy State Program Guide (Clean Energy Group and Smart Power 2009) provides information about marketing clean energy programs. The guide begins with a general introduction to marketing principles. Six marketing strategies are then described, with examples of successful implementation and action plans for each strategy. The report concludes with advice on how to begin developing a customized marketing plan.
    - The Distributed Renewable Energy Finance and Policy Toolkit (Kubert and Sinclair 2009) provides detailed descriptions of types of financial support that public benefit funds can provide to encourage the adoption of renewable energy technologies. The report describes each financing tool, outlines important strengths and weaknesses of the tool, lists best practices for successful implementation, and gives one or more examples of where the tool has been used.
- Database of State Incentives for Renewables & Efficiency (DSIRE) (North Carolina State University 2010) is a website funded by the U.S. Department of Energy and maintained by the North Carolina Solar Center and the Interstate Renewable Energy Council. DSIRE includes a complete listing of all renewable energy and efficiency incentives by state (including state-, local-, and utility-funded incentive programs), maps, tables, and a publications library. DSIRE is an excellent resource for in-depth research into the incentive programs in other states.

Employment Forecasts

Researchers at DOE, national and international agencies, and universities are studying the potential jobs associated with renewable energy in general and the PV products sector specifically. The following are a few especially relevant resources that will provide greater insight into the potential for solar PV job creation in the years to come.

- NREL's Jobs and Economic Development Impact (JEDI) model allows users to analyze the job and economic impact of power generation products, including biofuels, coal, concentrating solar power, natural gas, and wind. The user inputs such variables as location, year, type, size, and installed cost. The output is a summary of jobs resulting from both construction of the facility and long-term operation. Earnings and economic impact are included in the summary (http://www.nrel.gov/analysis/jedi/).
- NREL is also working on a forthcoming labor market analysis to support DOE's SunShot Initiative and to improve the JEDI model with respect to PV. The study includes primary-source research targeting installers and construction firms to identify workforce skills seen as lacking in the current labor pools (Friedman 2010a).
- The Solar Foundation's "National Solar Jobs Census 2010" targeted the remaining portions of the PV value chain. The study measured solar jobs in R&D, professional services, manufacturing, and wholesale trade, and provides insight into types of jobs, location, and growth potential. Secondary data for this study were derived from economic models based on Standard Industrial Classification (SIC) codes, while primary research was conducted through interviews of human resources personnel at various firms (Jordan/Green LMI 2010).

APPENDIX B: UNDERSTANDING THE PHOTOVOLTAIC PRODUCTS SECTOR

Polysilicon

Polysilicon manufacturing is a large, heavy-industrial chemical operation activity that is not suitable for most U.S. locations. It is a highly capital- and engineering-intensive activity, and it requires the investment of hundreds of millions of U.S. dollars. A nominal 5,000-ton annual capacity (roughly equivalent to 500 MW of c-Si panels per year) would require a construction period of 2 years, a long ramp-up period, hundreds of acres, and electrical connections in the tens of megawatts. Polysilicon manufacturing also benefits from "over-the-fence" access and proximity to other chemical plants for chlorine or caustic soda. Because they are so energy intensive, polysilicon plants often choose regions with exceptionally cheap power prices and low power price escalation assurances. Figure 12 is a photograph of solar-grade polysilicon feedstock.

A handful of major manufacturers produce almost all of the U.S.-made polysilicon. The leading producers are:

- Hemlock Semiconductor (United States)
- MEMC (United States)
- Mitsubishi (Japan and the United States)
- Renewable Energy Corporation (REC) (Norway)
- Tokuyama Corporation (Japan)
- Wacker-Chemie AG (Germany).

Photo by Warren Gretz, NREL/PIX 08791.

Figure 12. Solar-grade polysilicon feedstock.

Solar Glass

Solar glass manufacturing is a heavy-industrial operation. Solar glass manufacturing plants are typically capable of 200–600 tons of production per day, depending on process type and application—roughly equivalent to 700–1,200 MW of solar panels per year. Glass plants typically require 5–30 acres, and can require up to a few hundred million U.S. dollars in investment capital. Glass plants are also very energy intensive, burning natural gas to provide heat for production operations. Energy and raw materials make up more than 70% of the glass manufacturing cost of goods sold.

Because they are energy intensive and have large volume and weight of incoming materials and outgoing glass, glass plants often choose regions that have cheap energy prices and are closely connected to customers by rail.

Many container-glass and architectural-glass companies also produce solar glass, but they typically do so on different process equipment.

Bulk Chemicals and Gases

High-purity bulk gases like nitrogen, oxygen, helium, hydrogen, and argon are raw materials consumed later in the PV manufacturing process, often in large enough volumes to require on-site generation. Although these gases are not particularly hazardous or toxic, on-site generation equipment of these bulk gases can require significant site area and electric power. Gas plant equipment is often one of the tallest structures in a PV product manufacturing facility.

Bulk chemicals like hydrochloric, hydrofluoric, or nitric acid, as well as cleaning agents like potassium or sodium hydroxide, and surfactants also are often required in manufacturing PV products. A number of chemical companies supply these high-purity industrial chemicals.

Polysilicon, solar glass, and chemical and gas manufacturing have high barriers to entry due to high capital investment, large economies of scale, and mature technology and process expertise. These factors have kept the number of competitors for these markets in the United States low. A few major players in each major segment are competing in the U.S. market, and opportunities are somewhat limited to well-funded players with new differentiating technologies, or to expansions of existing players.

The Heart of the PV Products Sector: Ingot Preparation and Wafering, Cell Manufacturing, and Module Assembly

Most PV module production is based on c-Si materials and technology. The major manufacturing operations discussed here are for the c-Si strategy. Because of the growth in thin-film panels, however, information about thin film has been added where the panels are notably different.

c-Si Ingot Preparation and Wafer Slicing

In this process, a silicon ingot is grown from polysilicon and sliced into wafers. Ingot preparation and wafer slicing are typically done in separate buildings on the same site.

A nominal plant processes about 1 million kilograms (kg) of polysilicon per year (roughly 100 MW of annual capacity) and has typical investment requirements of about $50 million, although this can vary considerably depending on individual company strategies. Sites for ingot preparation and wafer slicing are typically greater than 5 acres and can be tens of acres. Ingot and wafer operations can require more than a 10-MW electrical connection, as well as large water and wastewater connections. Air emissions can include significant amounts of nitrogen oxide and particulates.

Ingot preparation and wafer slicing are not a distinct business for thin-film PV panels.

Cell Manufacturing

The second major step in the production of a PV panel is the transformation of a silicon wafer into a PV device or cell. This process typically takes place in a single building. Cell manufacturing processes have many variations that are tightly guarded by individual manufacturing companies as they define the performance of the PV product. For this reason,

PV cell manufacturing companies hold a significant portion of the PV industry's intellectual property.

A nominal cell manufacturing plant processes 120 MW of cells per year, requires tens of acres, and has a typical investment requirement of $40 million, although this can vary considerably depending on individual company strategies. Figure 13 shows a typical c-Si PV cell and part of an automated cell manufacturing line.

Cell manufacturing is usually not a distinct business for thin-film PV panels.

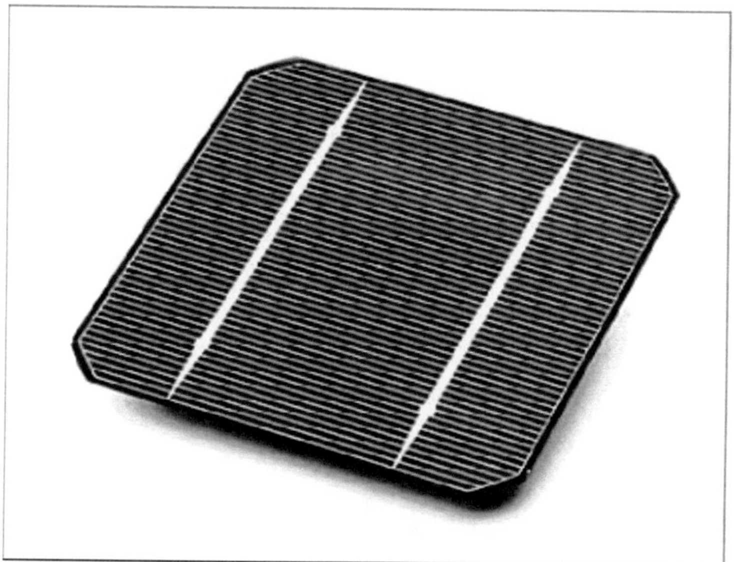

Photo by Rick Mitchell/PIX 04065.

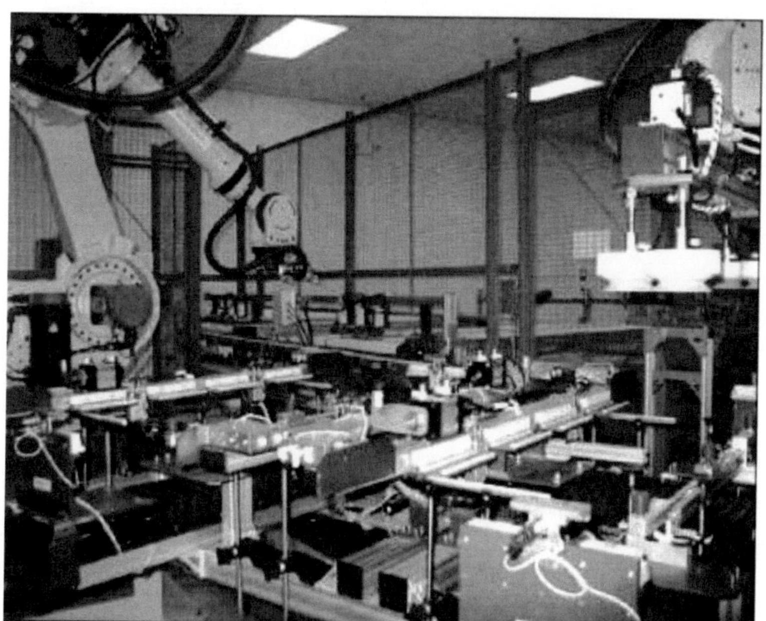

Photo by Shell Solar/PIX 13380.

Figure 13. Shell Solar Industries' crystalline silicon solar cell (left) and manufacturing robots (right).

Module Assembly

The final step in manufacturing a PV panel is module assembly, the process of connecting PV cells electrically and mechanically and then packaging them into usable PV panels (Figure 14). Typical packaging includes lamination and framing using glass and aluminum as support structures. Module assembly typically takes place in a single building and has relatively low infrastructure requirements.

A nominal module assembly manufacturing operation plant that processes 120 MW of cells per year would require tens of acres of site area and have an investment requirement of about $40 million, although this can vary considerably depending on individual company strategies.

In the United States, module assembly operations frequently are based on turn-key equipment provided by one of a few equipment companies. This equipment is generally highly automated— a different business model than is frequently used in Asia, where many operations are done by hand.

Module assembly is not a distinct step in the process of manufacturing thin-film PV panels.

Photo by BP Solar/PIX 13459.

Figure 14. BP Solar's crystalline silicon modules for residential and commercial buildings.

Thin Film and Other PV Panel Manufacturing Approaches

Crystalline silicon PV, which accounted for approximately 85% of PV module production in 2009, is not the only available manufacturing technology.

Thin-film PV typically uses thin layers of semiconductor materials like amorphous silicon, copper indium diselenide (CIS), copper indium gallium diselenide (CIGS), or cadmium telluride (CdTe). These PV module materials and manufacturing technologies produce PV modules that generally have lower efficiencies, but also lower cost per watt, than c-Si PV modules. Thin-film processes typically are more engineering-intensive and complex than c-Si processes. This is because there are fewer turn-key providers of manufacturing equipment, there is less industry experience, and the existing experience is split among the three major materials.

Thin-film manufacturing processes, unlike c-Si, are not easily split into multiple factories, producing cells in one location and modules in another. Because thin-film manufacturing operations generally span multiple materials and processes, they can also span a wide range of site selection or economic development criteria. In general, however, thin-film manufacturing is more equipment- and automation-intensive than are c-Si processes, and more focused than c-Si on intellectual property and research and development (R&D). Figure 15 shows a flexible thin-film product, as well as roll-to-roll deposition equipment. Flexible thin film is not the dominant thin-film PV product (most thin-film PV modules look very similar to c-Si modules).

Photo by UNI-SOLAR/PIX 15779.

Photo by Global Solar Energy/PIX 13415.

Figure 15. Thin-film laminate at UNI-SOLAR (left) and roll-to-roll deposition equipment at Global Solar Energy (right).

Inverters

Inverters are electrical products that convert the direct current (DC) electric output of solar modules into alternating current (AC) electricity. Grid-tied inverters, the market share leader, also integrate (characteristics such as frequency and phase) the energy into the existing electrical grid. PV inverters are typically divided into stand-alone, grid-tied, or battery-backup types, with some crossover. In the United States, however, grid-tied systems are the dominant type. Grid-tied inverters typically are designed specifically for residential, commercial, or utility applications.

U.S. inverter manufacturing businesses typically have a wide range of business models, manufacturing various components locally or outsourcing a range of components. U.S. inverter assembly facilities do not have especially high needs for any specific utility, specialized workforce, or other economic development requirement. They can also scale up and down relatively easily.

There are dozens of U.S. inverter companies, many specializing in specific technologies, at specific scales, or in serving local markets.

Photo by U.S. Postal Service/PIX 11016.

Figure 16. Commercial-scale PV system inverter.

Racking

Racking product companies typically produce support structures for fixed-tilt or tracking PV installations. Racking systems can also be specialized according to their intended use in ground or roof attachment.

Racking structures are primarily metal, based on standard angles, pipe, and flat steel and aluminum bar products, and assembled through standard metalworking processes like cutting, drilling, grinding, welding, and fastening. Racking also serves to route wiring and provide some grounding connections. The tracking system can include electromechanical assembly components.

Racking product manufacturing facilities vary widely, but generally they do not require large infrastructure connections. They are similar in appearance to warehouse or industrial space.

Figure 17 shows the physical structure of a two-axis PV tracking system holding eight PV modules. This tracker includes sensors and electromechanical assemblies, and it is one of the most complex examples of PV racking products.

Photo by SunPower/PIX 14598.

Figure 17. PV panels on a single-axis tracking array.

Site Selection Criteria *Land and Buildings*

Land and building requirements for manufacturing facilities vary greatly depending on product type and technology. Generally, land needs to be:

- Able to meet the development schedule, with required permits, approvals, title transfers, environmental studies, and utilities/infrastructure in place, and low development fees
- Appropriately zoned for proposed operating uses, with no code restrictions or easements that would affect operations; near compatible surrounding uses
- Flat in a square or rectangular configuration (ideally)

- Close to good transportation, including highways (and, in some cases, rail) for the shipping and receiving of large products.
- Different manufacturing processes require different types of buildings. Sometimes companies can take advantage of existing buildings to reduce schedule and startup costs. In most cases, it is the costs associated with infrastructure and permits (not the structure) that create advantages to building reuse.

Typical building requirements where building reuse is possible include the following:

- Building size requirements vary greatly by segment and can range from tens of thousands of square feet to several hundred thousand square feet in a single building.
- Buildings generally should be single story; if multiple story, they should have good connectivity between floors. Clear interior heights are driven by equipment requirements, but they can be above 20 feet. Floors must be able to support heavy industrial loads, without column spacing less than 30 feet.
- Existing infrastructure should be capable of meeting electrical, water and wastewater, and communication requirements, as well as occupancy codes and permits.
- Buildings should have no unmitigated hazardous material use or history.
- Buildings should have good truck access and adequate truck loading and parking spaces.

Utilities

Utility requirements of the PV products sector vary by product and technology. Typical utility requirements include:

- *Electricity.* Requirements range from conventional light industry loads to tens of megawatts for integrated panel manufacturing, glass, or polysilicon production.
- *Water and wastewater.* Polysilicon, silicon ingot preparation and wafer slicing, and cell manufacturing can require more than 1 million gallons of water per day. Many manufacturing processes use process chemicals and gases that may not meet discharge standards for existing wastewater treatment facilities.
- *Natural gas.* Standard industrial gas connection sizes, with the exception of solar glass manufacturing.
- *Communications.* Standard commercial voice and data based on office staff counts.

Available utilities should satisfy these criteria:

- *Competitive cost.* The cost of electricity is critical for large users such as polysilicon, solar glass, ingot preparation and wafer slicing, and cell manufacturing. Electrical costs at or below \$.06 to \$.08 per kWh (and, in extreme cases, as low as \$.03 to \$.04 per kWh) are ideal, and can include long-term contracts at fixed prices. Other segments of the PV industry have much lower electrical requirements and can tolerate higher electricity costs.
- *Sufficient supply.* Overall capacity and connections to the site without significant cost to the user, and ability to negotiate cost-sharing arrangements quickly.

- *Availability within the project schedule.* If system improvements are required, the area must be able to meet milestones and deadlines in the facility startup schedule.
- *Reliable and knowledgeable providers.* Local utilities should be familiar with the particular business and have experience in dealing with large or unique users.

Operating Cost Environment

PV products companies must reduce costs continually to remain competitive. The primary components of operating costs for PV products business vary. Generic c-Si and thin-film module manufacturing operating costs are shown in Tables 9 and 10, respectively. In the tables below, the sales price for thin film is set at 90% of the sales price of c-Si. While this is not universally true, it is a reasonable assumption that allows a comparison of the major cost categories of each manufacturing process.

Table 9. Crystalline Silicon PV Generic Costs

Crystalline Silicon PV Manufacturing - 100 MW	Total CAPEX ($)	Total Annual OPEX ($)	$/Watt	% of Cost/Watt
Ingot Preparation and Wafering:				
Factory Capital Investment (Deprec. Over 7 yrs)	$ 60,000,000	$ 8,900,00	$ 0.09	4%
Polysilicon		46,200,000	0.46	23%
Wire Saw Materials		9,800,00	0.10	5%
Ingot Casting Materials		7,300,000	0.07	4%
Labor		6,100,000	0.06	3%
Other Materials and Overhead Costs		8,800,000	0.09	4%
Ingot Preparation and Wafering Costs		87,100,00	0.87	44%
Cell Manufacturing:				
Factory Capital Investment (Deprec. Over 7 yrs)	40,000,000	5,600,000	0.06	3%
Labor		7,200,000	0.07	4%
Utilities		5,000,000	0.05	2%
Overhead Costs		10,808,000	0.10	5%
Materials (Spray on Etches, Screen Printing Inks, etc.)		12,000,000	0.12	6%
Cell Costs		39,880,000	0.40	20%
Module Assembly:				
Factory Capital Investment (Deprec. Over 7 yrs)	23,000,000	3,220,000	0.03	2%
Labor		4,000,000	0.04	2%
Overhead Costs		4,000,000	0.04	2%
Materials (Glass, EVA, Tedlar, Junction Box, Frame)		40,000,000	0.40	20%
Module Costs		51,220,000	0.51	26%
Total Annual Manufacturing Costs		178,200,000	1.78	0.89010989%
Gross Margin			0.22	11%
Total Panel Sales Price			2.00	100%

CAPEX = Capital expenditures.
OPEX = Operating expenditures.

Of the generic costs in Tables 9 and 10, the line items that vary the most by location are land and building costs, utilities, labor costs, and other overhead.

Table 10. Thin-Film PV Generic Costs

Thin Film (a-Si base) PV Manufacturing - 65 MW	Total CAPEX ($)	Total Annual OPEX ($)	$/Watt	% of Cost/Watt
Building (Depreciation over 15 years)	$ 50,000,000	$ 3,333,333	$ 0.05	3%
Tools (Depreciation over 7 years)	140,000,000	20,000,000	0.31	15%
Glass (Front and Back)		21,962,000	0.34	17%
Gasses and Chemicals		13,206,000	0.20	10%
Maintenance		9,391,000	0.14	7%
Utilities		9,000,000	0.14	7%
Framing Materials		8,000,000	0.12	6%
Labor		6,000,000	0.09	5%
Other Overhead		12,000,000	0.18	9%
Total Annual Manufacturing Costs		102,892,333	1.58	79%
Gross Margin			0.22	12%
Total Panel Sales Price			1.80	100%

CAPEX = Capital expenditures.
OPEX = Operating expenditures.

Tax structures and incentives vary greatly by location and can affect land and building costs and other overhead. Most state and local tax systems include a combination of income, sales, and property taxes that can affect each industry segment differently. There is no perfect tax system for all businesses, so most locations provide credits and other incentives to help overcome negative impacts to targeted companies. Major considerations can be how sales are apportioned for income tax purposes, whether there is a personal property tax, and how tax credits, abatements, and other incentives are applied.

Utilities vary significantly across the country according to power generation sources, utility structure, and taxes.

Schedule Considerations

Delivery contracts, incentives, and other deadlines make manufacturing project schedules important. Many PV product manufacturing facilities are large and require equipment that can have long lead times or build out periods. To meet these schedules, PV products sector companies look for the following:

- Land and buildings that are "shovel ready" (with clear ownership, immediate availability at a listed price, and a transferable title) and that meet infrastructure requirements
- Multiple appropriate sites and alternatives if the preferred site or facility falls through
- Zoning, permits, and other regulatory and environmental approvals that are complete or in progress, funded, and approved in writing
- Required infrastructure that is ready or in progress, funded, and approved in writing (infrastructure primarily includes electricity, water and wastewater, and transportation)

- Staff members who are experienced in meeting schedules for similar projects, as well as experienced with governmental processes for occupancy codes, hazardous gases and chemicals, air emissions, and special building requirements
- Staff members who are experienced in writing a memorandum of understanding that defines a clear process for receiving incentives and development assistance
- A facility engineering, procurement, and construction company that has a successful local history and on-schedule, within-budget experience with similar facilities.

Transportation and Proximity to End Markets

PV products sector companies often are very shipping-intensive. Easy transportation access is important, especially for segments with large economies of scale.

Proximity to end-user markets in the PV products sector has not been a critical factor to date, but it is becoming more important as transportation costs increase. The products segment most influenced by proximity to end markets is module assembly.

Technology Support, Higher Education, and Research

Proximity to existing technology companies, higher education programs, and related research is a site selection criterion for PV products sector businesses whose advantage is in the intellectual property and not only the low manufacturing cost. Companies in the early stages of development will benefit the most from access to these resources. For these companies, the following are attractive area assets:

- An existing cluster of similar businesses. Clusters provide companies with an environment that includes existing technology R&D, technology workers, and focused education and training.
- A business culture of innovation. Product companies like to have access to other researchers and institutions to drive product development.
- Access to local universities. Universities provide science and engineering expertise that is important to manufacturing companies. They also supply employees, continuing education, and faculty members who may collaborate on research.

In addition to providing a supportive and attractive environment for growing or recruiting existing companies, these area characteristics create an environment for the spinoff and creation of new PV product companies.

Workforce and Training

Workforce requirements vary by segment. For most PV products sector businesses, only a small number of staff members are engineering or management professionals. Most workers are skilled or semi-skilled operators with experience in manufacturing, such as the following:

- Equipment operators
- Process technicians
- Maintenance technicians
- Line supervisors.

In addition to an available, trained workforce, PV products sector companies often look for a training system. The majority of PV products sector workers can receive training from community colleges and local technical schools.

Public and Private Support and Policies

Areas that have been successful in attracting PV products sector businesses have created an overall environment that is supportive of the PV industry in general. Although financial incentives are important public commitments, successful economic development strategies have also adopted a more comprehensive approach, including investment in the following:

- A regulatory environment that is supportive, easy to understand, and easy to navigate. This can include:
    - Public offices that understand PV products sector metrics and vocabulary
    - Regulatory processes that are understandable and reasonable
    - The support and encouragement of a local solar market.
- Long-term, visible, and proactive support for the PV products sector and a pro-business climate
- The involvement and support of local electric utilities; utility support and involvement in the form of special rates can be as important as public policy for some segments.
- Special districts and incentives to targeted PV products sector businesses.

Incentives

Successful local incentives target specific segments based on local goals. While federal incentives are critically important, they are available throughout the United States and are not viable regional or local economic development tools.

The most effective local incentives are financial. While incentives are one of the most powerful tools available to economic development groups, incentives alone normally are not enough for a company to make an investment decision. A few of the most common incentives are as follows:

- Cash grants or low-cost financing
- Tax credits, particularly those that can be monetized or passed through
- Tax holidays and abatements
- Free or low-cost land; waiver of development and other fees, and worker training costs; or reduced utility hookup costs
- Funding to participate in R&D partnerships
- Expedited permits and other regulatory processes
- Reduced utility costs
- Off-take agreements (agreements by a public or private entity to purchase a set amount of a PV products sector company's production over a certain period). Often, such an agreement provides the company with an order commitment that can be used to help secure project financing.

Key Lessons about the PV Products Sector

The PV products supply chain includes a variety of different manufacturing processes, each with very distinct requirements from land and energy-intensive chemicals production, from specialized cell manufacturing to less specialized assembly operations and production of inverters and racking. Companies may specialize in one small portion of the supply chain, or combine production steps in a single location.

Communities should understand each product in the supply chain and the requirements of each manufacturing process. It is important to be able to realistically evaluate the resources available in the community against the needs of the manufacturers when deciding what types of PV business to target in an economic development campaign.

Local incentives and improvements should be tailored to the needs of the targeted sectors to emphasize strengths and mitigate weaknesses. Communities should also be aware of other locations that could potentially compete for the same businesses, and monitor incentives and improvements undertaken by these competing areas.

Appendix C: Understanding the Photovoltaic Energy Generation Sector

Value Chain—from Components to Kilowatt-Hours

The following PV energy generation segments typically are increasingly distinguished at larger scales, specifically at the commercial and utility scales, where larger project budgets and complexity require specialized expertise. At residential scales, there are generally only a few project participants, most commonly the land or building owner and the installer/system integrator or system integrator in partnership with a third-party financier/owner, as illustrated in Figure 9.

As economic developers focus on developing attractive local PV installation markets to create local jobs, it is important to note in the following supply chain that the two major segments with the largest jobs impacts are the segments most directly involved in PV system installation:

- System or technology integration
- Engineering, procurement, and construction.

System or Technology Integrators

System integration is the process of intelligently matching and combining system components into an installation that is physically, aesthetically, and electrically designed to perform well at a given site.

System integrators must know the range of products, the performance specifications, and the requirements of the project site. Their services attempt to minimize the project's cost, maximize its energy production, and satisfy its technical requirements.

System integrators' facility needs are neither utility-intensive nor complex, and they typically use existing buildings. Some system integrators may require space for testing and simulation, training, or warehousing. System integration companies can be small and local, or large and national. At residential and small commercial scales, system integrators often do installation themselves. At larger commercial and utility scales, system integrators typically are not involved in installation labor.

Project Developers

For large commercial- and utility-scale projects, project developers find and create PV project opportunities. They then bring together the required partners to perform the work, often providing initial and high-risk capital. PV project developers typically hire service providers to propose and evaluate sites; model system performance; conduct environmental, land use, and interconnection studies and apply for permits; negotiate power purchase agreements (PPAs) with the utility; negotiate incentives with economic development groups; lobby local residents; and hire construction and operation providers.

Most important, a project developer's primary expertise is in the legal and financial structure of a PV project. This includes the energy, debt and equity, and energy credit transactions among all of the project stakeholders and regulatory agencies.

Developers may also provide a source of capital to fund projects, take initial risks such as paying for feasibility studies, and take on other, ongoing project risk.

Engineering, Procurement, and Construction Firms

At commercial and utility scales, engineering, procurement, and construction companies bundle together many of the technical services hired by the project developer into a single contract. They often begin with PV system performance modeling, conduct environmental and interconnections studies, and end with system commissioning. Their fundamental role in most PV projects is the design and construction; however, they may also provide schedule, cost, and performance guarantees. Some specialty firms will also take on project risk or participate in the project beyond its traditional role as a technical and construction services provider.

Engineering, procurement, and construction firms typically have direct design and project management staff, but use local subcontractors for construction and installation services. As a result, the facility needs usually are limited to office space and limited construction management space on project sites.

Engineering, procurement, and construction firms typically are hired through a request for proposals process on a project-by-project basis.

Operation and Maintenance Contractors

Commercial and utility PV projects have ongoing monitoring and maintenance requirements that typically are contracted to firms with expertise in performance monitoring and maintenance (such as panel washing and PV component replacement). These ongoing services are important and local, but they create very few ongoing jobs.

At residential and small commercial scales in the right climate, these activities are often the responsibility of the building or property owner. However, third-party ownership paired with subcontracts for performance monitoring and maintenance is also common.

Project Owner

Project owners can be land or building owners, or, as is common with third-party models, own only the solar equipment and lease the roof or land. Often, when the tax credits or depreciation and tax benefits have run their course, project ownership will change.

Electrical Utility or Power Purchaser

Increasingly, electric utilities are playing a role in PV energy generation, either as power purchasers (forming agreements with PV energy generation project owners or developers to buy electricity through PPAs) or as PV project owners and developers themselves. Utility structure and attitudes can play a large part in the overall PV energy generation climate. Supportive electric utilities often see PV energy generation sector growth as a way to satisfy renewable portfolio standards (RPS) goals, delay or eliminate pending investment in electrical infrastructure, or reduce peak loads. Utilities with this perspective may create PV energy generation incentives, and be economic development partners. Unsupportive electric utilities are important challenges that local economic developers need to recognize and work to overcome.

Renewable Energy Certificate Owner

RPSs are important drivers of the PV energy generation sector. RPS goals are typically state goals, met through the purchase or creation of renewable energy certificates (usually measured by the kilowatt-hour or megawatt-hour). In many cases, RECs are granted to project owners; however, RECs typically are sold through long-term agreements with electric utilities or power purchasers.

Site Selection Criteria

Table 11 highlights the major financial categories involved in building and operating a utility-scale PV energy generation project. The cost of utility-scale PV materials within the United States generally is not subject to much regional variation. Installation labor, however, can vary according to the degree of experience and competition available at a given site, and the degree of complexity driven by project conditions.

Solar Resources

Solar resources are not something that can be changed. Every city's solar resource potential is fixed, and this resource is an important criterion in defining the potential PV energy generation sector's evaluation of a city. Solar resources are typically measured in kilowatt-hours per square meter (m^2)/year or kilowatt-hours/m^2/day, and combined with other location-specific variables and into kilowatt-hour/kilowatt factors.

Experienced energy generation sector companies will know what a city's solar resource is, and know what this means for the local market. The general public, however, may not accurately understand the solar resource. In many cases, a cold or harsh climate is mistaken for a poor solar resource, and this is not an accurate assumption. The best and most widely used solar resource data comes from the National Renewable Energy Laboratory.

While solar resources are fixed, city economic developers should understand these resources to comprehend their competitive position and the gaps that must be bridged through incentives. Good solar resource understanding can also be the basis of marketing to combat public misperceptions about the viability of solar energy.

Table 11. Sample Utility-Scale PV Energy Generation Project Costs

Solar PV Farm - System Size = 50 MW	Total CAPEX ($)	Total Annual OPEX ($)	$/Watt	% of CAPEX Invt.
Modules CAPEX	$ 100,000,000		$ 2.00	65%
Rack (Fixed)	12,970,000		0.26	8%
Inverter	12,410,00		0.25	8%
Other BOS Materials	9,220,000		0.18	6%
Labor - Installation - Craft ($40/hr)	7,520,000		0.15	5%
Labor - CM & Engineering-Professional ($80/hr)	7,520,000		0.15	5%
EPC Expense	1,600,000		0.03	1%
Site Work Expense	1,600,000		0.03	1%
Other Indirect	940,000		0.02	1%
Land	940,000		0.02	1%
Total Balance of Plant	**54,720,000**		1.09	**35%**
Total Installed System CAPEX	**154,720,000**		3.09	**100%**
Total Annual Operating Costs (Labor & Expense)		$ 800,000	0.02	1%

Example Only

CAPEX = Capital Expenditures BOS = Balance of System EPC = Engineering, Procurement, and Construction
OPEX = Operating Expenditures BOP = Balance of Plant CM = Construction Management

Local Cost of Electricity and Utility Attitudes and Structures

PV energy generation sector projects ultimately compete with the local price of a kilowatt-hour delivered by the local electric utility. When a PV energy generation sector company selects project sites, it looks for the following:

- *A high competing cost of electricity.* The higher the competitive price, the easier it is to make a PV energy generation project viable or to realize a greater financial return.
- *An energy demand profile that aligns with PV energy generation.* If loads are driven by space cooling from solar heat gain (e.g., air conditioning) or other demands that align with peak solar generation, PV energy would be well suited to offset this energy requirement.
- *A PV-friendly utility.* Utility structure and attitude have a large influence on the attractiveness of a city for PV energy generation project development. Investor-owned utilities operate differently than cooperative utilities, and differently than municipal utilities. While one structure is not inherently more PV-friendly than another, cooperative utilities and municipal utilities have more history of PV energy generation project involvement. Investor-owned utilities need to balance the needs of their shareholdes with the desires of their customers, which may make it more difficult to support PV energy generation projects.

Public and Private Support

In many locations, PV energy generation sector projects require the financial and non-financial support of local governments to become viable in the short term. Such support can often overcome challenges identified in local solar resources or competing cost of electricity.

Overall community support can significantly affect PV market transformation. The leadership of public officials and local institutions can be influential.

Workforce and Training

While the range of PV energy generation sector services is broad, local economic developers are best served by recruiting PV energy generation projects, as project location (installation location) is the largest driver of local jobs.

The largest workforce requirement of PV energy generation projects is installation, construction, or trade labor. Training this workforce is an important step to building a market; however, many areas have existing and often available construction laborers who can easily be retrained.

Key Lessons about the PV Generation Sector

While there are a variety of pieces to the PV generation value chain, encouraging installation of generation projects is the primary focus of successful economic development efforts in this sector.

The solar resources of an area are somewhat important to the potential success in developing the PV generation sector (more so for encouraging utility-scale installations), but communities should not assume that cold weather makes an area unsuitable for PV generation. As discussed above and illustrated in the case studies in Section 3, many states in the northern United States and Canada have had great success in encouraging installation of PV systems and increasing the size of the PV generation workforce and the economic development opportunities from PV generation businesses.

The cost of conventional energy and attitudes of local utilities toward solar and other alternative energy sources are of paramount importance to the success of the PV generation industry. Public and private support in the form of policy, financial, and non-financial support can encourage the market and often compensate for factors such as low competing energy costs and less-than-ideal solar resources.

NOTICE

This report was prepared as an account of work sponsored by an agency of the United States government. Neither the United States government nor any agency thereof, nor any of their employees, makes any warranty, express or implied, or assumes any legal liability or responsibility for the accuracy, completeness, or usefulness of any information, apparatus, product, or process disclosed, or represents that its use would not infringe privately owned rights. Reference herein to any specific commercial product, process, or service by trade name, trademark, manufacturer, or otherwise does not necessarily constitute or imply its endorsement, recommendation, or favoring by the United States government or any agency thereof. The views and opinions of authors expressed herein do not necessarily state or reflect those of the United States government or any agency thereof.

ACKNOWLEDGMENTS

This report was prepared for the U.S. Department of Energy Solar Energy Technologies Program by CH2MHill.

This report was made possible through funding from the U.S. Department of Energy's Solar America Communities program. To learn more, please visit www.solaramerica communities.energy.gov.

SOLAR AMERICA COMMUNITIES

BIBLIOGRAPHY

American Solar Energy Society and Management Information Services, Inc. (2008). *Defining, Estimating, and Forecasting the Renewable Energy and Energy Efficiency Industries in the U.S. and in Colorado*. December. http://www.ases.org/images/stories/ASES/pdfs/CO_Jobs_Final_Report_December2008.pdf. Accessed on September 27, 2010.

Bartlett, J.E.; Margolis, R.M.; and Jennings, C.E. (2009). *The Effects of the Financial Crisis on Photovoltaics: An Analysis of Changes in Market Forecasts from 2008 to 2009*. National Renewable Energy Laboratory Technical Report, NREL/TP-6A2-46713. http://www.nrel.gov/docs/ fy10osti/46713.pdf. Accessed on September 28, 2010.

Clean Energy Group and Smart Power. (2009). *Smart Solar Marketing Strategies: Clean Energy State Program Guide*. August. http://www. cleanegroup.org/Reports/CEG_Solar_Marketing_Report_August2009.pdf. Accessed on October 14, 2010.

Environmental Defense Fund and Ella Baker Center for Human Rights. (2008). *Green Jobs Guidebook: Employment Opportunities in the New Clean Economy.* http://www.edf.org/cagreenjobs. Accessed on September 28, 2010.

Friedman, B. (2009). *NREL PV Jobs/Labor Intensity Project: New Ideas in Educating a Workforce in Renewable Energy and Energy Efficiency*. November 19. http://irecusa.org/wpcontent/uploads/2009/ 11/Friedman. pdf. Accessed on September 27, 2010.

Friedman, B./ (2010a). National Renewable Energy Laboratory Senior Energy Analyst. Personal communication with Amy Maule/CH2M HILL. July 14.

Friedman, B./ (2010b). Personal communication with Amy Maule/CH2M HILL. August 16.

Glass Association of North America, Flat Glass Manufacturing Division. (date?). *Float Glass Manufacturing Process Video.* http://www. glasswebsite.com/video/fgmd.asp. Accessed on October 27, 2010.

Global Insight, Inc. (2008). *U.S. Metro Economies, Current and Potential Green Jobs in the U.S. Economy.* October.____http://www.usmayors.org /pressreleases/uploads/ greenjobsreport.pdf.Accessed on September 27, 2010.

Goldberg, M.; Cliburn, J.K. (2011). *Economic Impacts from the Boulder County, Colorado ClimateSmart Loan Program using Property Assessed Clean Energy Financing.* MGA & Associates. January. Soon to be available on the NREL publications database.

Greenpeace and European Photovoltaic Industry Association. (2007). *Solar Generation IV – 2007: Solar electricity for over one billion people and two million jobs by 2020*.http://www.greenpeace.org/international/en/ publications/reports/solar-generation-iv/. Accessed on September 28, 2010.

Interstate Renewable Energy Council. (2010). *Education Information.* http://irecusa.org/irec-programs/workforce-development/education-information/. Accessed on November 30, 2010.

Jordan, P./Green LMI. (2010). Personal Communication with Amy Maule/CH2M HILL. August 19.

Kubert, C.; Sinclair, M. (2009). *Distributed Renewable Energy Finance and Policy Toolkit.* Clean Energy States Alliance. December. http://www. cleanenergystates.org/ Publications/cesafinancial_Toolkit_Dec2009.pdf. Accessed on September 28, 2010.

Lantz, E.; Tegen, S. (2009). *NREL Response to the Report Study of the Effects on Employment of Public Aid to Renewable Energy Sources from King Juan Carlos University (Spain).* White Paper, NREL/TP-6A2-46261. http://www.nrel.gov /docs/fy09osti/46261.pdf. Accessed on September 27, 2010.

Mehta, S. (2009). *PV Manufacturing in the United States: Market Outlook, Incentives and Supply Chain Opportunities*. Executive Summary. GTM Research. September.

Network for New Energy Choices, Vote Solar Initiative, and Interstate Renewable Energy Council. (2010). *Freeing the Grid: Best Practices in State Net Metering Policies and Interconnection Procedures.* http://www.newenergychoices.org/ uploads/Freeing TheGrid2010.pdf. Accessed on December 9, 2010.

North Carolina State University. (2010). *Database of State Incentives for Renewables & Efficiency: Summary Maps*. http://www.dsireusa.org/ summarymaps/index.cfm?ee= 1&RE=1. Accessed on September 28, 2010.

O'Rourke, S.; Kim, P.; Polavarapu, H. (2009). *Solar Photovoltaic Industry: Looking Through the Storm*. Deutsche Bank Securities Inc. January 21.

Roberts, B. (2008a). *Photovoltaic Solar Resource of the United States*. National Renewable Energy Laboratory. http://www.nrel.gov/gis/images. Accessed on January 31, 2011.

Roberts, B. (2008b). *Wind Resource (50m) of the United States*. National Renewable Energy Laboratory. http://www.nrel.gov/gis/images map_ wind_national_lo-res.jpg. Accessed on January 31, 2011.

Schreiber, D. (2010). *Hot PV Markets in the U.S.* Presented at Intersolar West, San Francisco. July 14.

Sherwood, L. (2009). *U.S. Solar Market Trends 2008*. Interstate Renewable Energy Council. July. http://irecusa.org/fileadmin/user_ upload/National OutreachDocs/SolarTrends Reports/IREC_Solar_Market_Trends_ Report_2008.pdf. Accessed on November 30, 2010.

Solar Energy Industry Association. (2004). *Our Solar Power Future. The US PV Industry Roadmap through 2030 and beyond.*

State of California, Department of Industrial Relations. (2010). *Division of Apprenticeship Standards - California Needs Apprentices in Green & Growing Fields*. http://www. dir.ca.gov/das/DAS_ ApprenticesGreen GrowingFields/DAS_ Apprentices GreenGrow ingFields.htm. Accessed on September 28, 2010.

State of New Jersey, Department of Labor and Workforce Development. (2010). *Incentives & Training Programs: Apprenticeship Training—Office of Grants Operations.*

http://lwd.dol.state.nj.us/labor/employer/ training/Apprenticeship.html. Accessed on September 28, 2010.

Sterzinger, G.; Svrcek, M. (2005). *Solar PV Development: Location of Economic Activity*. Renewable Energy Policy Project, Technical Report. January. http://www.repp.org/Solar_PV_Dev.htm. Accessed on September 28, 2010.

U.S. Department of Energy, Energy Efficiency and Renewable Energy. (2010). *2008 Solar Technologies Market Report.* DOE/GO-102010-2867. January. http://www1.eere.energy.gov/solar/pdfs/46025.pdf. Accessed on October 28, 2010.

U.S. Department of Energy, Energy Efficiency and Renewable Energy and Solar America Cities. (2009). *Solar Powering Your Community: A Guide for Local Governments*. July. http://www.solaramericacities.energy. gov/ resources/guide_for_local_governments/. Accessed on October 14, 2010.

U.S. Department of Energy, National Renewable Energy Laboratory. (2005). *Best Research-Cell Efficiencies*. http://www.nrel.gov/pv/thin_film/ docs/ kaz_best_research_cells.ppt. Accessed on September 28, 2010.

U.S. Department of Energy, National Renewable Energy Laboratory. (2009). *The Effects of the Financial Crisis on Photovoltaics: An Analysis of Changes in Market Forecasts from 2008-2009*. http://www.nrel.gov/ docs/fy10osti/46713.pdf. Accessed on September 28, 2010.

U.S. Department of Energy, National Renewable Energy Laboratory. (2010). *Dynamic Maps, GIS Data, & Analysis Tools: Solar Maps.*_http://www.nrel.gov/gis/solar.html. Accessed on November 30, 2010.

U.S. Department of Energy, National Renewable Energy Laboratory. (2010a). *JEDI: Jobs and Economic Development Impact Models*. http://www. nrel.gov/analysis/jedi/. Accessed on September 28, 2010.

U.S. Department of Energy, National Renewable Energy Laboratory. (2010b). Photographic Information eXchange. http://www.nrel.gov/data/pix/. Accessed on October 19, 2010.

U.S. Energy Information Administration. (2010). *Energy Outlook 2010, High and Low Economic Growth Tables*. Report #: DOE/EIA-0383(2010). http://www.eia.gov/oiaf/ieo/index.html. Accessed on September 28, 2010.

Wei, M.; Patadia, S.; Kammen, D.M. (2010). "Putting renewables and energy efficiency to work: How many jobs can the clean energy industry generate in the U.S.?" *Energy Policy*, 38 (2010) 991-931. http://rael. berkeley. edu/node/585. Accessed on December 2, 2010.

Wiser, R.; Barbose, G.;Peterman, C. (2009). *Tracking the Sun, The Installed Cost of Photovoltaics in the U.S. from 1998-2007*. LBNL-1516E. February 27. http://eetd.lbl.gov/ea/ems/reports/lbnl-1516e.pdf. Accessed on September 28, 2010.

Worldwatch Institute. (2008). *Green Jobs: Towards Decent Work in a Sustainable, Low-Carbon World.* September. http://www.unep. org/ labour_environment/features/greenjobs.asp. Accessed on September 27, 2010.

End Notes

[1] Scale-based definitions of these market segments vary throughout the solar industry. These system size ranges are defined strictly for the purposes of this report.

[2] For more information about financial tools, see Kubert and Sinclair 2009, Mehta 2009, and North Carolina State University 2010.

[3] For more information about financial tools, see also Kubert and Sinclair 2009; Mehta 2009; and North Carolina State University 2010.

In: Photovoltaic
Editors: Eric Falchuk and Carol Woodlee

ISBN: 978-161942-045-8
© 2012 Nova Science Publishers, Inc.

Chapter 2

TRACKING THE SUN III: THE INSTALLED COST OF PHOTOVOLTAICS IN THE U.S. FROM 1998-2009*

Galen Barbose, Naïm Darghouth and Ryan Wiser

EXECUTIVE SUMMARY

As the deployment of grid-connected solar photovoltaic (PV) systems has increased, so too has the desire to track the installed cost of these systems over time and by location, customer type, system characteristics, and component. This report helps to fill this need by summarizing trends in the installed cost[1] of grid-connected PV systems in the United States from 1998 through 2009 (updating two previous reports with data through 2007 and 2008, respectively), and providing preliminary cost trends for systems installed in 2010.[2] The analysis is based on installed cost data for approximately 78,000 residential and non-residential PV systems, totaling 874 megawatts (MW) and representing 70% of all grid-connected PV capacity installed in the United States through 2009.

Key findings of the analysis are as follows: [3]

- The capacity-weighted average installed cost of systems completed in 2009 – in terms of real 2009 dollars per installed watt (DC-STC)[4]
  incentives or tax credits – was \$7.5/Watt, virtually unchanged from 2008, and \$0.3/W below the averages in 2006 and 2007. From 1998-2009, capacity-weighted average installed costs declined by about 3.2% (or \$0.3/W) per year, on average, starting from \$10.8/W in 1998.
- Preliminary cost data suggest that a significant decline in average installed costs occurred in 2010, with the average cost of systems installed through the California Solar Initiative (CSI) program declining by \$1.0/W during the first ten months of

---

* This is an edited, reformatted and augmented version of an Environmental Energy Technologies Division, Lawrence Berkeley National Laboratory publication, dated December 2010.

2010 relative to 2009, and the average cost of systems installed in New Jersey declining by $1.2 /W during the first six months of 2010 compared to 2009.

- Installed costs lagged wholesale PV module price movements from 2007 -2009. Over this period, wholesale PV module prices declined by $1 .3/W (based on Navigant Consulting's Global Power Module Price Index), while total installed costs declined by only $0.2/W. The preliminary 2010 cost data cited above, however, suggests that the drop in wholesale module prices during the preceding years translated into a large reduction in installed costs in 2010.
- Both module and non-module costs have declined significantly over time. From 1998-2009, wholesale module prices dropped by $1 .9/W (40%), while from 1998-2007, implied non- module costs (which may include such items as inverters, mounting hardware, labor, permitting and fees, shipping, overhead, taxes, and installer profit) fell by $2.5/W (40%).[5]
- The year-on-year trend in average installed cost varied among system size ranges. From 2008 to 2009, the average cost of systems in the 5-10 kW and 10-100 kW size ranges declined by $0.4/W and $0.3/W, respectively, but remained flat for systems <5 kW and 100- 500 kW in size, and rose by $0.3/W for systems >500 kW.
- The distribution of installed costs within a given system size range narrowed significantly from 1998 to 2005, with high-cost outliers becoming increasingly infrequent, indicating a maturing market. However, little if any further narrowing of the cost distribution occurred from 2005 through 2009.
- PV installed costs exhibit significant economies of scale, with systems ≤2 kW completed in 2009 averaging $9.9/W, while >1,000 kW systems averaged $7.0/W (or about 29% less). Two multi-MW, utility-scale PV systems installed in 2009 that are not included in the primary data sample, but for which cost data were obtained through public sources, had significantly lower installed costs ($2.5/W and $5. 1/W) than the >1,000 kW systems in the primary data sample.
- Component-level cost data for systems installed in 2009 indicate that module and inverter costs were relatively invariant across system sizes, while the remaining set of other costs were $0.5-0.6/W lower for 10-100 kW systems than for either smaller or larger systems. This contrasts with trends from the preceding two years, when average module costs were lower for larger systems.
- International experience suggests that greater near-term cost reductions may be possible with increased market scale in the United States, as the average installed cost of 3-5 kW residential PV installations in 2009 (excluding sales/value-added tax) was significantly lower in both Germany ($4.7/W) and Japan ($5.9/W) than in the United States ($7.7/W).
- Average installed costs vary widely across states; among ≤10 kW systems completed in 2009, average costs range from a low of $7.1/W in Texas to a high of $9.6/W in Minnesota. This variation in average installed cost across states suggests that, in addition to absolute market size, other state and local factors (e.g., permitting requirements, labor rates, and sales tax exemptions) strongly influence installed costs.
- The average installed cost of residential systems installed in 2009 was lower than that of other similarly sized systems. Compared to commercial systems, for example,

residential systems had average installed costs that were lower by approximately $0.5/W within both the 5-10 kW and 10-100 kW size ranges. In contrast, public sector systems installed in 2009 had relatively high average installed costs, exceeding the average installed cost of commercial systems by $0.7/W for systems in the 10-100 kW size range and by $0.8/W for systems in the 100-500 kW size range.

- The new construction market offers significant cost advantages for residential PV; among 1- 3 kW residential systems funded through two California programs (the New Solar Home Partnership Program and the California Solar Initiative) and installed in 2009, PV systems installed in residential new construction cost $1 .6/W less than comparably-sized residential retrofit systems (or $1 .9/W less if focused exclusively on rack-mounted systems).
- PV systems with thin-film modules had higher average installed costs in 2009 than comparably-sized crystalline systems ($0.8/W higher among ≤10 kW systems and $0.4/W less among 10-100 kW systems).
- Average installed costs were lowest for systems with mid-range module efficiencies in 2009. Among systems ≤10 kW, for example, systems with module efficiencies of 15-16% had an average installed cost of $7.4/W, compared to $8.2/W for systems with module efficiencies ≤12% and $8.3/W for systems with module efficiencies >18%.
- Among ≤10 kW systems installed in 2009, those with tracking (either single- or double-axis) had average installed costs $1.7/W (or 21%) higher than fixed-axis systems (both roof- mounted and ground-mounted).[6] This cost differential may reflect both the additional cost of tracking equipment, as well as the increased cost associated with ground-mounting, which is more prevalent among tracking systems than fixed-axis systems.[7]
- The average cash incentive provided by the state/utility PV incentive programs in the sample ranged from $1.2-$2.2/W for systems installed in 2009, depending on system size. This represents up to a $0.3/W decline from 2008 and a $2.0-$3.3/W decline from the peak in 2002.
- In 2009, the average combined after-tax value of state/utility cash incentives plus state and federal investment tax credits (ITCs) – but excluding revenue from the sale of renewable energy certificates or the value of accelerated depreciation – was $3.9/W for both residential and commercial PV.8 This represents roughly a 37% increase for residential PV relative to2008, and is a consequence of the lifting of the dollar cap on the federal residential ITC for residential systems installed after January 1, 2009. In contrast, the combined after-tax incentive for commercial PV in 2009 remained effectively unchanged from the year prior and was slightly below its peak of $4.2/W in 2006.
- In 2009, the average net installed cost faced by PV system owners – that is, installed cost minus after-tax incentives – stood at $4.1/W for residential PV and $4.0/W for commercial PV. For residential PV, this represents an historic low, and is $1 .3/W (or 24%) below the 2008 average, reversing the trend of the preceding several years during which average net installed costs had been slowly rising. For commercial PV, the average net installed costs in 2009 was virtually unchanged from the preceding

two years, and was up slightly (by $0.4/W or 10%) from the all-time low of $3.6/W in 2006.

- Financial incentives and net installed costs diverged widely across states in 2009. Among residential PV systems completed in 2009, the combined after-tax incentive ranged from an average of $3.5/W in California to $5.9/W in New York, and net installed costs ranged from an average of $2.4/W in Texas to $5.5/W in Minnesota. Incentives and net installed costs for commercial systems varied similarly across states.

1. Introduction

Installations of solar photovoltaic (PV) systems have been growing at a rapid pace in recent years. In 2009, approximately 7,500 megawatts (MW) of PV were installed globally, up from approximately 6,000 MW in 2008, consisting primarily of grid-connected applications.[9] With 435 MW of grid-connected PV capacity added in 2009, the United States was the world's fourth largest PV market in 2009, behind Germany, Italy, and Japan.[10] Despite the significant year-on-year growth, however, the share of global and U.S. electricity supply met with PV remains small, and annual PV additions are currently modest in the context of the overall electric system.

The market for PV in the United States is driven by national, state, and local government incentives, including up-front cash rebates, production-based incentives, requirements that electricity suppliers purchase a certain amount of solar energy, and federal and state tax benefits. These programs are, in part, motivated by the popular appeal of solar energy, and by the positive attributes of PV – modest environmental impacts, avoidance of fuel price risks, coincidence with peak electrical demand, and the possible deployment of PV at the point of use. Given the relatively high cost of PV, however, a key goal of these policies is to encourage cost reductions over time. Therefore, as policy incentives have become more significant and as PV deployment has accelerated, so too has the desire to track the installed cost of PV systems over time, by system characteristics, by system location, and by component.

To address this need, Lawrence Berkeley National Laboratory initiated a report series focused on describing trends in the installed cost of grid-connected PV systems in the United States. The present report, the third in the series, describes installed cost trends from 1998 through 2009, and provides preliminary cost data for systems installed in 2010.[11] Importantly, in focusing on *installed cost*, this report does not address other factors that may also influence the *levelized cost of electricity* from PV, such performance improvements and financing costs.

The analysis is based on project-level cost data from approximately 78,000 residential and nonresidential PV systems in the U.S., all of which are installed at end-use customer facilities (herein referred to as "customer-sited" systems). [12] The combined capacity of systems in the data sample totals 874 MW, equal to 70% of all grid-connected PV capacity installed in the United States through 2009 and representing one of the most comprehensive sources of installed PV cost data for the U.S.[13] The report also briefly compares recent PV installed costs in the United States to those in Germany and Japan. Finally, it should be noted that the analysis presented here focuses on descriptive trends in the underlying data, serving primarily to summarize the data in tabular and graphical form; later analysis may explore some of these trends with more-sophisticated statistical techniques.

The report begins with a summary of the data collection methodology and resultant dataset (Section 2). The primary findings of the analysis are presented in Section 3, which describes trends in installed costs prior to receipt of any financial incentives: over time and by system size, component, state, system ownership type (customer-owned vs. third party-owned), host customer segment (residential vs. commercial vs. public-sector vs. non-profit), application (new construction vs. retrofit), and technology type (building-integrated vs. rack-mounted, crystalline silicon vs. thin- film, and tracking vs. fixed-axis). Section 4 presents additional findings related to trends in PV incentive levels over time and among states (focusing specifically on state and utility incentive programs as well as state and federal tax credits), and trends in the net installed cost paid by system owners after receipt of such incentives. Brief conclusions are offered in the final section, and several appendices provide additional details on the analysis methodology and additional tabular summaries of the data.

2. DATA SUMMARY

This section briefly describes the procedures used to collect, standardize, and clean the data provided by individual PV incentive programs, and summarizes the basic characteristics of the resulting dataset, including: the number of systems and installed capacity by PV incentive program; the sample size relative to all grid-connected PV capacity installed in the United States; and the sample distribution by year, state, and project size.

Data Collection, Conventions, and Data Cleaning

Requests for project-level installed cost data were sent to state and utility PV incentive program administrators from around the country, with some focus (though not exclusively so) on relatively large programs. Ultimately, 27 PV incentive programs provided project-level installed cost data from 16 states. [14] To the extent possible, this report presents the data as provided directly by these PV incentive program administrators; however, several steps were taken to standardize and clean the data, as briefly summarized here and described in greater detail in Appendix A.

In particular, two key conventions used throughout this report deserve specific mention:

1. All cost and incentive data are presented in real 2009 dollars (2009$), which required inflation adjustments to the nominal-dollar data provided by PV programs.
2. All capacity and dollars-per-watt ($/W) data are presented in terms of rated module power output under Standard Test Conditions (DC-STC), which required that capacity data provided by several programs that use a different capacity rating be translated to DC-STC.[15]

The data were cleaned by eliminating projects with clearly erroneous cost or incentive data, by correcting text fields with obvious errors, and by standardizing identifiers for module and inverter models. To the extent possible, each PV system in the dataset was classified as either building- integrated PV or rack-mounted and as using either crystalline or thin-film modules, based on a combination of information sources. Finally, data on market sector (e.g., residential, commercial, government, non-profit) were not provided for roughly 40% of the systems in the final dataset; for the purpose of calculating the value of state and federal investment tax credits and net installed costs, systems ≤10 kW were assumed to be residential, and systems >10 kW were assumed to be commercial, if not identified otherwise.[16]

Table 1. Data Summary by PV Incentive Program

State	PV Incentive Program	No. of Systems	Total MWDC	% of Total MWDC	Size Range (kWDC)	Year Range
AZ	APS Solar & Renewables Incentive Program	2,291	15.0	1.7%	0.4 - 255	2002 - 2009
	SRP EarthWise Solar Energy Program	1,039	7.4	0.9%	0.7 - 502	2005 - 2009
CA	CEC Emerging Renewables Program (ERP)	27,947	146.4	16.8%	0.1 - 670	1998 - 2008
	CEC New Solar Homes Partnership (NSHP)	2,597	8.3	0.9%	1.2 - 154	2007 - 2009
	CPUC California Solar Initiative (CSI)	25,074	326.0	37.3%	1.1 - 1,308	2007 - 2009
	CPUC Self-Generation Incentive Program (SGIP)	806	148.6	17.0%	34 - 1,266	2002 - 2009
	LADWP Solar Incentive Program	2,199	22.9	2.6%	0.3 - 1,176	1999 - 2009
	SMUD Residential Retrofit and Commercial PV Programs	368	2.5	0.3%	1.2 - 172	2005 - 2009
CT	CCEF Onsite Renewable DG Program	117	13.6	1.6%	1.6 - 570	2003 - 2009
	CCEF Solar PV Program	829	4.9	0.6%	0.8 - 19	2005 - 2009
FL	FEO Solar Energy System Incentives Program	577	4.6	0.5%	2.0 - 1,016	2006 - 2009
MA	MassCEC Small Renewables Initiative*	1,990	20.3	2.3%	0.2 - 502	2002 - 2009
MD	MEA Solar Energy Grant Program	546	2.2	0.2%	0.5 - 45	2005 - 2009
MN	MSEO Solar Electric Rebate Program	198	0.8	0.1%	0.5 - 40	2002 - 2009
NH	NHPUC Renewable Energy Rebate Program	189	0.5	0.1%	0.4 - 5.0	2008 - 2009
NJ	NJCEP Customer Onsite Renewable Energy Program	3,859	74.5	8.5%	0.8 - 2,372	2004 - 2009
	NJCEP SREC Registration Program	134	26.9	3.1%	0.7 - 1,650	2007 - 2009
	NJCEP Renewable Energy Incentive Program	641	5.6	0.6%	0.6 - 50	2009 - 2009
NV	NPC/SPPC RenewableGenerations Rebate Program	499	2.9	0.3%	0.4 - 99	2004 - 2009
NY	NYSERDA PV Incentive Program	1,990	13.9	1.6%	0.7 - 79	2003 - 2009
OR	ETO Solar Electric Program	1,321	12.3	1.4%	0.4 - 859	2003 - 2009
PA	DEP Sunshine Solar PV Program (and other state agency programs**)	372	2.8	0.3%	1.1 - 123	2009 - 2009
	SDF Solar PV Grant Program	164	0.7	0.1%	1.2 - 12	2002 - 2008
TX	Austin Energy Power Saver Program	1,072	4.8	0.5%	0.2 - 28	2004 - 2009
	IOU Solar Incentive Programs (AEP, Entergy, Oncor, SWEPCO, TNMP)	154	1.2	0. 1%	0.7 - 236	2009 - 2009
VT	RERC Small Scale Renewable Energy Incentive Program	365	1.3	0.2%	0.2 - 38	2004 - 2009
WI	Focus on Energy Renewable Energy Cash-Back Rewards Program	614	3.3	0.4%	0.5 - 47	2002 - 2009
Total		*77,952*	*874.1*	*100%*	*0.1 - 2,372*	*1998 - 2009*

* Included within the totals shown for MassCEC's Small Renewables Initiative are systems funded through predecessor PV incentive programs offered by the Massachusetts Technology Collaborative.

** Pennsylvania state agencies have offered various grant and rebate programs for renewable energy systems; for simplicity, these programs are summarized in aggregate. Of these programs, the largest funding source for PV systems through 2009 has been the Department of Environmental Protection (DEP)'s Sunshine Program.

Sample Description

The final dataset, after all data cleaning was completed, consists of approximately 78,000 grid- connected, residential and non-residential PV systems, totaling 870 MW (see Table 1). 17 This represents approximately 70% of all grid-connected PV capacity installed in the United States through 2009, and about 63% of 2009 capacity additions (see Figure 1). The largest state markets missing from the primary data sample, in terms of cumulative grid-connected PV capacity installed through 2009, are: Colorado (4.7% of total U.S. installed capacity), Hawaii (2.1%), and North Carolina (1.0%).

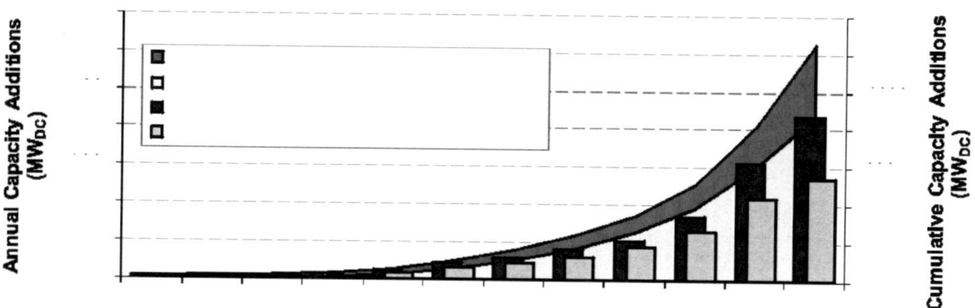

Data source for cumulative and annual U.S. grid-connected PV capacity additions: Sherwood, L. 2010. U.S. Solar Market Trends 2009. Interstate Renewable Energy Council.

Figure 1. Data Sample Compared to Total U.S. Grid-Connected PV Capacity.Table 2. Data Sample by Installation Year.

Table 2. Data Sample by Installation Year

	1998	1999	2000	2001	2002	2003	2004	2005	2006	2007	2008	2009	Total
No. of Systems	39	180	217	1,308	2,478	3,474	5,589	5,587	8,684	12,635	14,108	23,653	77,952
% of Total	<0.5%	<0.5%	<0.5%	2%	3%	4%	7%	7%	11%	16%	18%	30%	100%
Capacity (MWDC)	0.2	0.8	0.9	5.4	15	33	45	62	90	130	219	272	874
% of Total	<0.5%	<0.5%	<0.5%	1%	2%	4%	5%	7%	10%	15%	25%	31%	100%

The primary sample consists only of data provided by PV incentive program administrators, which consist entirely of customer-sited systems. The report separately describes the installed cost of 11 multi-MW grid-connected PV systems not included in the primary dataset, including the three largest PV systems installed in the United States through 2009.[18]Cost data for these projects were compiled from press releases and other publicly available sources. The data for these 11 projects bring the total PV capacity for which cost data are presented to 982 MW, equal to 78% of all grid- connected PV capacity installed in the United States through 2009.

The PV systems in the primary dataset were installed over a twelve-year period, from 1998 through 2009. Given the dramatic expansion of the U.S. solar market in recent years, however, the sample is skewed towards projects completed during the latter years of the study period, with approximately half of the PV systems and more than half of the total capacity in the sample installed in 2008 and 2009 (see Table 2). See Appendix B for annual installation data (number of systems and capacity) disaggregated by PV incentive program and by system size range.

Among the 27 PV incentive programs from which data was provided, the great majority of the sample is associated with the four largest PV incentive programs in the country to-date: California's Emerging Renewables Program (ERP); California's Self-Generation Incentive Program (SGIP); the California Solar Initiative (CSI) Program; and New Jersey's Customer Onsite Renewable Energy (CORE) Program. As such, the sample is heavily weighted towards systems installed in California and New Jersey, as shown in Figure 2. In terms of installed capacity, these two states represent75% and 12% of the total data sample, respectively. Arizona, Massachusetts, Connecticut, Oregon, and New York each represent 1.6-2.6% of the sample, with the remaining nine states (Florida, Nevada, New Hampshire, Maryland, Minnesota, Pennsylvania, Texas, Wisconsin, and Vermont) comprising 2.9% in total.

The size of the PV systems in the primary dataset spans a wide range, from as small as 100 W to as large as 2.3 MW, but almost 90% of the projects in the sample are ≤10 kW (see Figure 3). In terms of installed capacity, however, the sample is considerably more evenly distributed across system size ranges, with systems >100 kW comprising 47% of the total installed capacity, and systems ≤10 kW comprising 35%.

3. PV Installed Cost Trends

This section presents the primary findings of the report, describing trends in the average installed cost of grid-connected PV based on the dataset described in Section 2. It begins by presenting the trends in installed costs over time; by system size; by component; between the United States, Germany, and Japan; among individual states; between customer-owned and third party-owned systems; and among customer types (residential, commercial, public sector, and non-profit).[19]

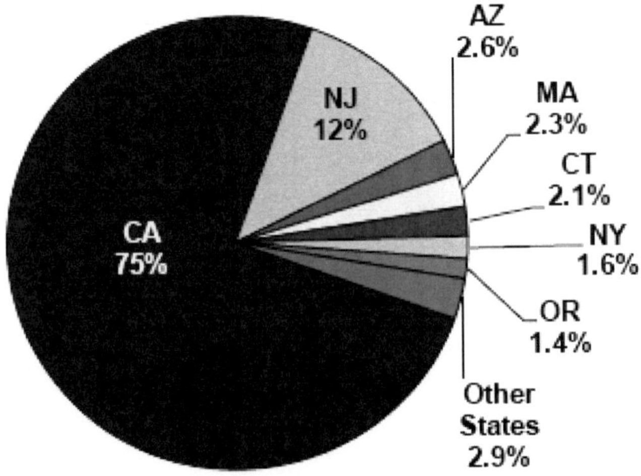

Figure 2. Data Sample Distribution among States (by Cumulative MW).

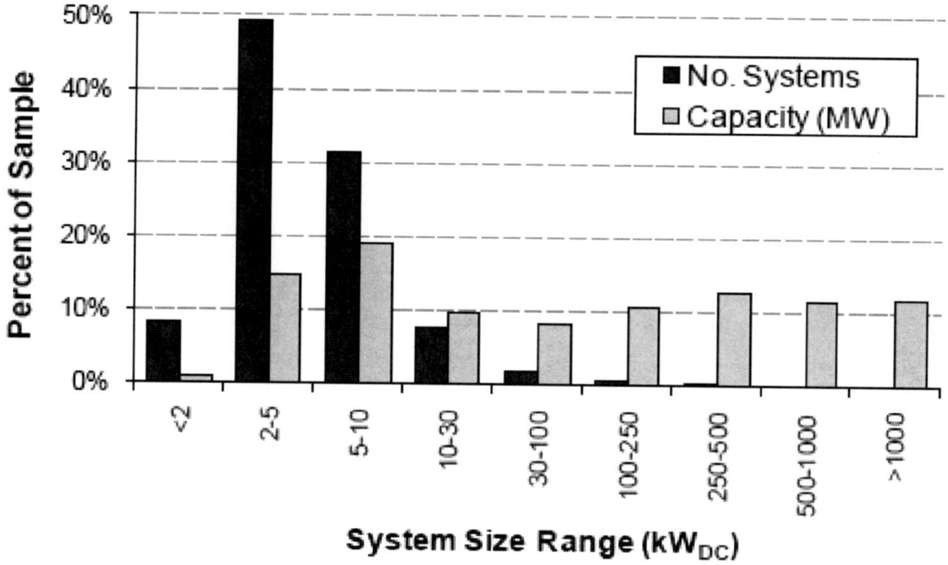

Figure 3. Data Sample Distribution by PV System Size.

It then compares installed costs among several specific types of applications and technologies – specifically, residential new construction vs. residential retrofit, BIPV vs. rack-mounted systems, systems with thin-film modules vs. those with crystalline modules, and tracking vs. fixed-axis systems. To be clear, the focus of this section is on installed costs, as paid by the system owner, prior to receipt of any financial incentives (e.g., rebates, tax credits, etc.).

Overall Average Installed Costs Remained Flat from 2008 to 2009

Figure 4 presents the average installed cost of all projects in the primary sample completed each year from 1998-2009. Over the entirety of the twelve-year period depicted in the figure, capacity- weighted average installed costs declined by a total of $3.3/W (30%), or $0.3/W (3.2%) per year, on average. This cost decline occurred primarily during the period 1998-2005, after which average costs declined by only a modest amount ($0.2/W total from 2005-2009).

Focusing specifically on the cost trend from 2008 to 2009, the capacity-weighted average annual installed cost for the entire data sample remained unchanged at $7.5/W, while the simple (arithmetic) average installed cost fell by $0.2/W, from $8.2/W to $8.0/W. This trend stands in stark contrast to the widely reported decline in wholesale module prices during 2009. One potential cause for this discrepancy is that the cost of many projects installed in 2009, as paid by the final system owner, may have been based on contracts signed and inventory stocked at a time when wholesale module prices were higher than at the date of system installation. This hypothesis is consistent with the preliminary 2010 cost data presented in Text Box 1, which show that average installed costs declined significantly from 2009 to 2010.

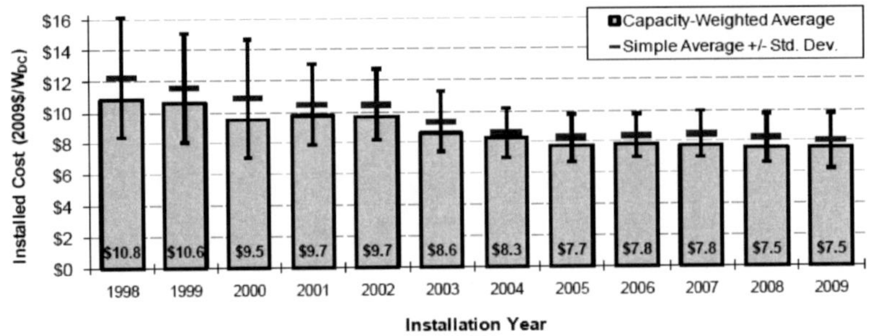

Notes: See Table 2 for sample size by installation year.

Figure 4. Installed Cost Trends over Time.

Text Box 1. Preliminary Installed Cost Trends for 2010

The dramatic and widely reported decline in *wholesale* module prices that began in 2008 and continued through 2009 and into 2010 would suggest that *retail* installed costs for PV systems should decline as well. Although the data presented in Figure 4 show that installed costs did not noticeably decline from 2008 to 2009, preliminary cost data for projects installed in 2010 indicate a significant drop in average installed costs between 2009 and 2010.

Figure 5 compares the average cost of projects installed in 2009 to the cost of projects installed during the first 6-10 months of 2010, with results presented separately for California's CSI program, based on data through November 10, 2010, and New Jersey, based on data from all statewide incentive programs through June 30, 2010. Among the California systems, the capacity-weighted average cost of all projects installed during the first ten months of 2010 was $1 .0/W below the average in 2009. Similarly, for New Jersey, the capacity-weighted average installed cost of systems installed during the first six months of 2010 was $1 .2/W below the average of all systems installed in 2009.

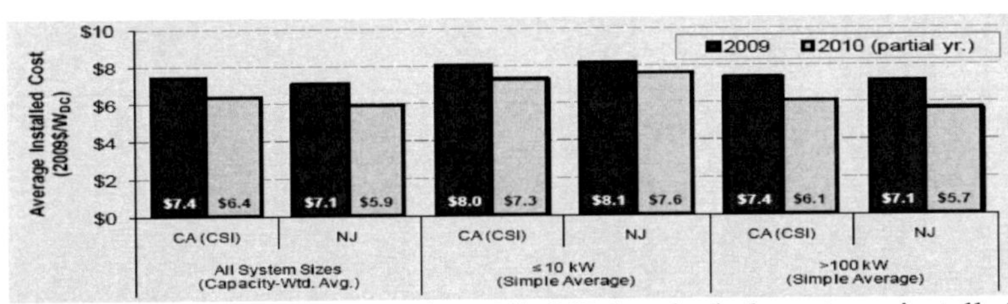

Notes: CA data are for the CSI program only, while NJ data include systems installed through the CORE Program, Renewable Energy Incentive Program, and SREC Registration Program. The 2010 partial year data extend through November 10, 2010 for CA systems and through June 30, 2010 for NJ systems.

Figure 5. 2009 and Preliminary 2010 Installed Costs for California and New Jersey.

In both California and New Jersey, the decline in average installed costs was significantly greater for large systems than for smaller systems. Specifically, among CSI systems in California, average costs dropped by $1 .3/W for systems >100 kW and by $0.7/W for systems ≤10 kW. Similarly, in New Jersey, average costs dropped by $1.4/W for systems >100 kW and by $0.5/W for systems ≤10 kW.

A Decline in Average Module Costs from 2008 to 2009 Was Offset by an Apparent Increase in Non-Module Costs

Figure 6 disaggregates average annual installed costs into average module and non-module costs. Because many programs did not provide component-level cost data, this figure relies on Navigant Consulting's Global Power Module Price Index, which reflects wholesale module prices. In order to roughly approximate the lag between movements in wholesale module prices and retail installed costs for PV systems, the module price index values shown in Figure 6 are equal to the average of the current and prior years' Global Power Module Price Index.[20] The non-module costs shown in Figure 6 (which may include such items as inverters, mounting hardware, labor, permitting and fees, shipping, overhead, taxes, and installer profit) were then calculated as the difference between the average total capacity-weighted installed cost and the "lagged" module price index in each year. Given the approximations inherent in this method, caution is warranted in interpreting these results.

Notwithstanding that caveat, Figure 6 clearly shows that both module and non-module costs declined significantly over the twelve-year period from 1998-2009, with module costs falling by \$1.9/W and non-module costs falling by \$1.4/W. Within the last two years of the data series, from 2008-2009, module costs (which, again, represent the average of the current and prior years' wholesale index price) declined by \$0.7/W.[21] In contrast, non-module costs, as calculated according to the method described above, *increased* by approximately \$0.7/W from 2008-2009.

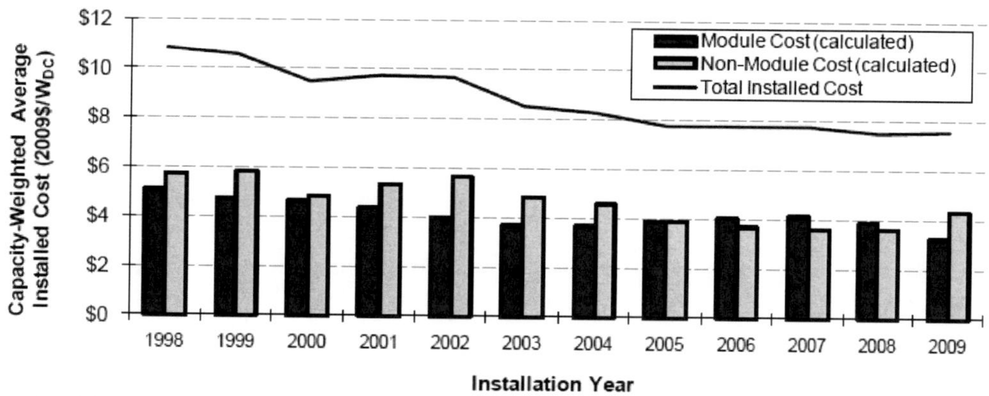

Notes: Module costs are equal to the average of the current and prior years' global module price index, developed by Navigant Consulting. Non-module costs are calculated as the reported total installed costs minus the averaged module index.

Figure 6. Module and Non-Module Cost Trends over Time.

This apparent increase in non-module cost may, in part, reflect increased installer profit margins, capacity constraints in the delivery infrastructure, or higher penetrations of tracking systems or emerging technologies that increase installed costs while improving system performance. It is also possible, however, that the method used to calculate module and non-module costs for Figure 6 does not fully account for the lag between wholesale module price movements and PV system installations, and could therefore overstate the increase in non-module costs from 2008-2009.Trends in non-module costs may be particularly relevant in gauging the impact of state and utility PV programs. Unlike module prices, which are primarily established through global markets, non-module costs consist of a variety of cost components that may be more readily affected by local programs – including deployment programs aimed at increasing demand (and thereby increasing competition and efficiency among installers) as well as more-targeted efforts, such as training and education programs. Thus, the fact that non-module costs have fallen over time, at least until 2005, suggests that state and local PV programs may have had some success in driving down the installed cost of PV. At the same time, the relatively flat trend in non-module costs from 2005-2009 suggests either that state and local programs have become less effective at driving down non-module costs, or that other drivers (e.g., constraints in the delivery infrastructure, permitting and incentive application costs, or installer profit margins) have put increasing upward pressure on non-module costs borne by PV system owners.

Average Installed Costs Fell Moderately in 2009 For Systems 5-100 kW, While Remaining Flat or Rising for Other System Sizes

Although average installed costs remained relatively flat across all size ranges from 2008 to 2009, small differences can be observed among the size classifications shown in Figure 7.[22] Specifically, average costs declined by $0.4/W and $0. 3/W, respectively, for systems in the 5-10 kW and 10-100 kW size ranges, but remained flat for systems <5 kW and 100-500 kW in size, and rose by $0.3/W for systems >500 kW. As noted previously in Text Box 2, however, preliminary cost data for 2010 indicates that significant cost reductions occurred between 2009 and the first six to ten months of 2010, with the greatest cost reductions occurring among larger systems.

Over the entirety of the historical analysis period, cost reductions are most evident for smaller system sizes. From 1999-2009, the average installed cost of systems ≤5 kW dropped by $3 .4/W, from $1 1.8/W to $8.4/W, while the average cost of systems in the 5-10 kW and 10-100 kW size ranges dropped by $2.2/W and $3. 1/W, respectively. It is less apparent to what extent larger systems experienced cost reductions over the 1999-2009 period, given limited data availability for the early years of the analysis period. Over the historical periods for which sufficient data is available, the average cost of systems 100-500 kW declined by $1 2/W (2002-2009), while the average cost of systems >500 kW increased by $0.2/W (2003-2009).[23]

The Distribution of Installed Costs Narrowed from 1998 to 2005, but No Further Narrowing Occurred through 2009

As indicated by the standard deviation bars in Figure 4, the distribution of installed costs has narrowed considerably over time. This trend can be seen with greater precision in Figure 8 and Figure 9, which present frequency distributions of installed costs for systems less than and greater than 10 kW, respectively, installed in different time periods.

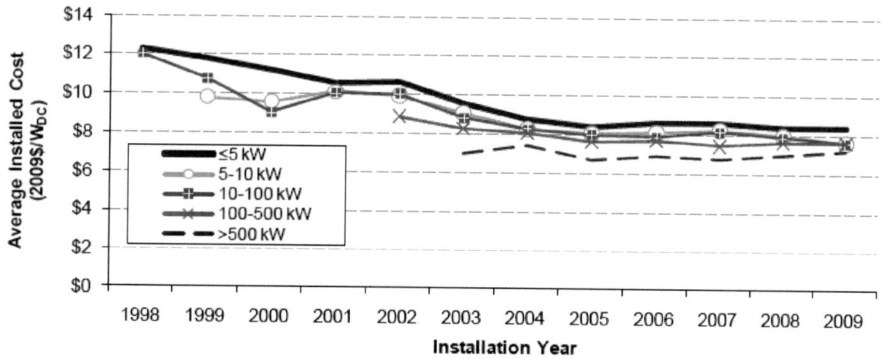

Notes: Averages shown only if five or more observations were available for a given size category in a given year. For sample sizes, refer to Table B-2 in the Appendix.

Figure 7. Installed Cost Trends over Time, by PV System Size.

Both figures show a marked narrowing of the cost distributions occurring between 1998 and 2005, although this trend largely subsided from 2005 through 2009. This convergence of prices, with high-cost outliers becoming increasingly infrequent, is consistent with a maturing market characterized by increased competition among installers and module manufacturers and by better-informed consumers.

The two figures also show a *shifting* of the cost distributions to the left, as would be expected based on the previous finding that average installed costs have declined over time. For example, in both figures, the cost distributions for 2008-2009 are shifted to the left relative to the distributions for 2006-2007. This is consistent with Figure 4, which shows that the simple average (i.e., arithmetic average) installed cost in 2008 and 2009 was lower than in 2006 and 2007.

Installed Costs Exhibit Economies of Scale

Large PV installations may benefit from economies of scale through price reductions on volume purchases of materials and the ability to spread fixed costs and transaction costs over a larger number of installed watts. This expectation has generally been borne out in experience, as indicated by Figure 10, which shows the average installed cost according to system size, for PV systems completed in 2009.

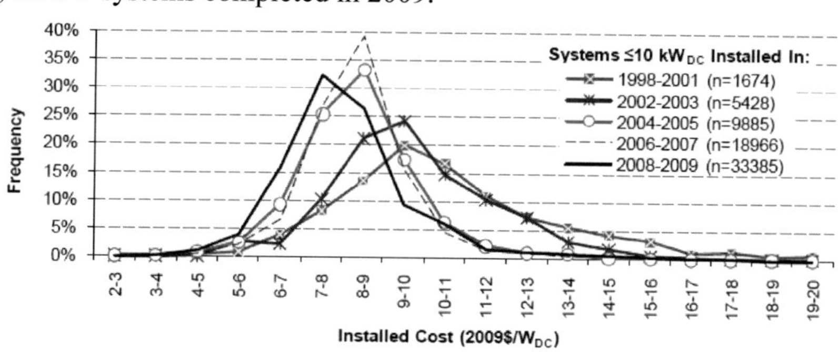

Figure 8. Distribution of Installed Costs for Systems ≤10 kW.

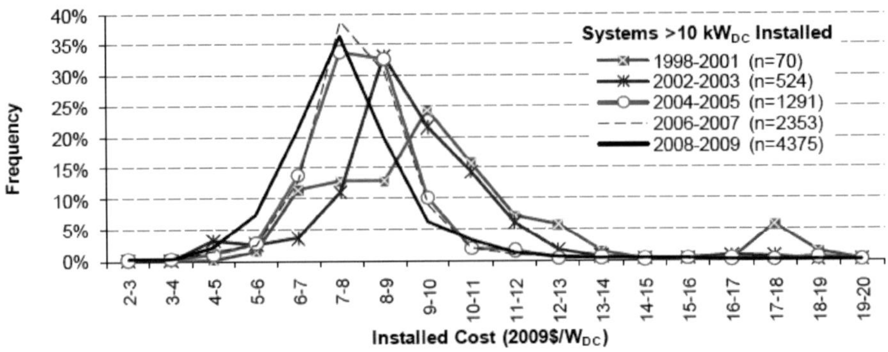

Figure 9. Distribution of Installed Costs for Systems >10 kW.

The smallest systems (≤2 kW) exhibit the highest average installed cost ($9.9/W), while the largest systems (>1000 kW) have the lowest average cost ($7.0/W, or about 29% below the average cost of the smallest systems). Interestingly, economies of scale do not appear to be continuous with system size, but rather, most strongly accompany increases in system size up to 10 kW, and increases in system size above 250 kW. In contrast, the data do not show evidence of significant economies of scale within the 30-250 kW size range, and in fact, they indicate that average installed costs *increase* somewhat with system size between 10 kW and 250 kW. This latter trend may reflect a lower level of standardization as system size increases and/or increased permitting costs, which could counteract scale economies within certain system size ranges.

The primary dataset underlying the results shown in Figure 10 consists only of data provided by the 27 PV program administrators in our sample. Not included in this dataset are a number of large, multi-MW PV systems, several of which are installed on the utility-side of the meter. Installed cost data for 11 of these projects have been reported in press releases and other public sources, and are summarized in Table 3.[24] As shown, the installed costs of these projects vary considerably. The two projects completed in 2009 have reported installed costs of $5.1/W and $2.5/W, which are significantly below the average for systems >1,000 kW shown in ($7.0/W).

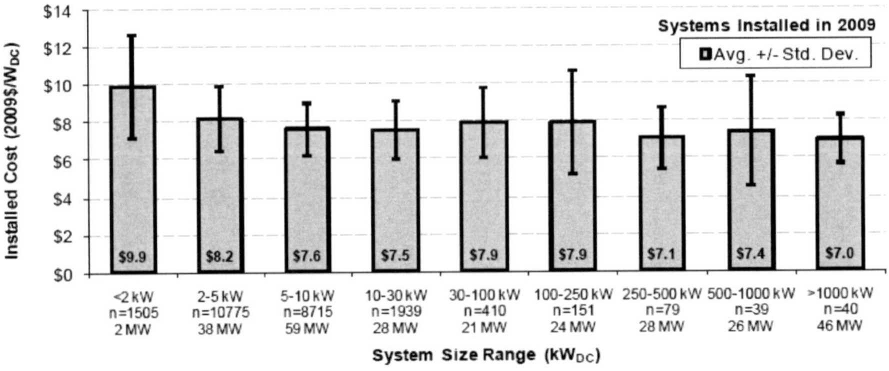

Figure 10. Variation in Installed Cost According to PV System Size.

Also note that a number of the systems in Table 3 have tracking systems, and are therefore likely to attain higher performance (and thus lower *levelized* costs on a $/MWh basis, even if the up-front installed costs are higher) than the large projects in the primary dataset, which are mostly fixed-axis systems.

To the extent that the economies of scale described above have persisted over time, they partially explain the temporal decline in average installed costs highlighted earlier, as the average size of PV systems has grown over time. As shown in Figure 11, the average size of ≤10 kW systems grew from 2.7 kW in 1998 to 4.7 kW in 2009, while the average size of systems >10 kW rose from 25 kW to 65 kW over the same time period. [25]

Non-Module/Non-Inverter Costs Were Lowest for Mid-Sized Systems in 2009

The average module and non-module costs presented in Figure 6 were estimated based on a wholesale module price index. This approach was necessitated, in part, by the fact that many of the PV incentive programs in our data sample did not provide component-level cost data.

Table 3. Installed Cost of Large (≥ 2 MW) Out-of-Sample PV Systems

Location	Year of Installation	Plant Size (kWDC)	Installed Cost (2009$/WDC)	Tracking System Design
Arcadia, FL	2009	30,000	5.1	single axis
Blythe, CA	2009	25,200	2.5	none (fixed-axis)
Boulder City, NV	2008	12,600	3.2	none (fixed-axis)
Fairless Hills, PA	2008	3,000	6.6	none (fixed-axis)
Fontana, CA	2008	2,400	4.2	none (fixed-axis)
Riverside, CA	2008	2,000	6.5	none (fixed-axis)
Nellis, NV	2007	14,200	7.3	single axis
Alamosa, CO	2007	8,220	7.6	fixed, single axis, and double axis
Fort Carson, CO	2007	2,000	6.5	none (fixed-axis)
Springerville, AZ	200 1-2004	4,590	6.1	none (fixed-axis)
Prescott Airport, AZ	2002-2006	3,388	5.6	single axis and double axis

Notes: Cost for Springerville is for capacity added in 2004. Cost for Prescott is for single-axis capacity additions in 2004.

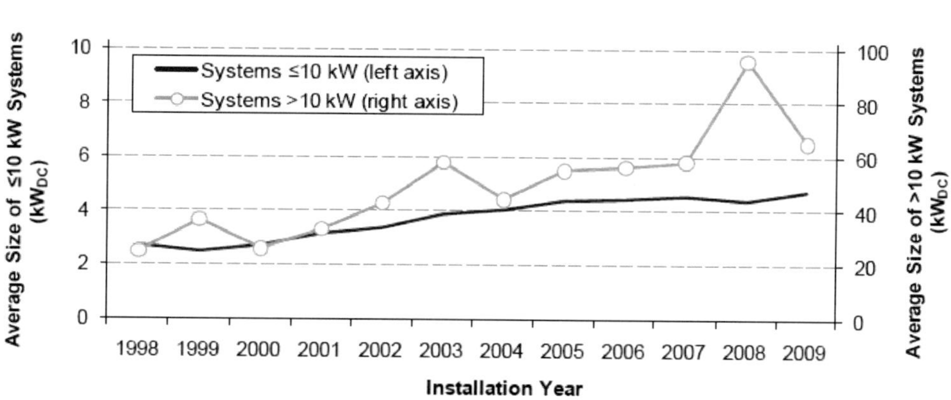

Figure 11. PV System Size Trends over Time.

However, three programs[26] did provide component-level cost data (even if at a fairly coarse level of detail), and these data provide a moderate level of additional detail on the composition of total system costs and the variation in component-level costs across system sizes.

Figure 12 summarizes the component-level cost data provided by PV incentive programs in our data sample, for systems installed in 2009. As shown, average module costs ranged from $4.1/W to $4.2/W, representing 52% to 56% of total installed costs, across the three size ranges. On average, inverter costs comprised 6-9% of the total cost, while other costs (e.g., mounting hardware, labor, overhead, installer profit, etc.) made up the relatively substantial remaining 36-42%.

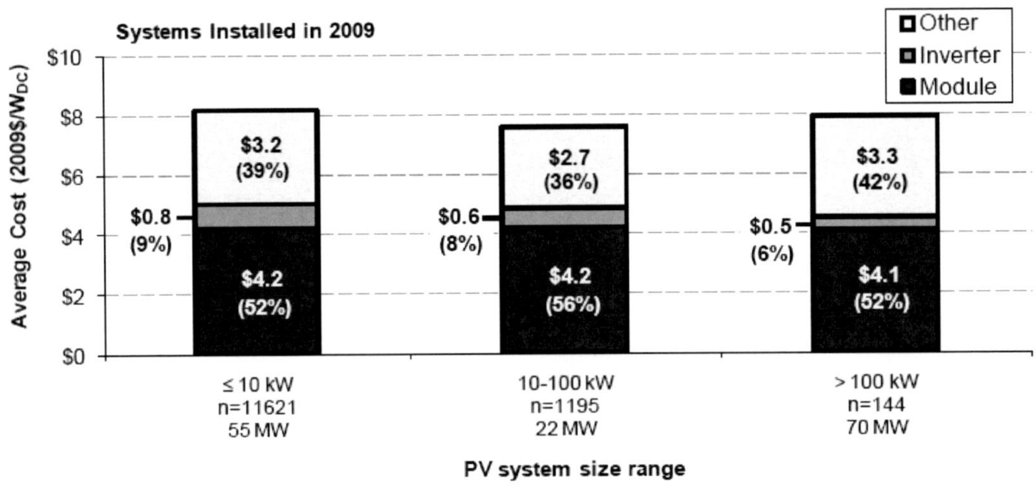

Figure 12. Module, Inverter, and Other Costs.

Comparing across system size ranges, Figure 12 indicates that average module costs were relatively invariant across system size in 2009. This contrasts with the preceding two years, as shown in Table 4, when reported module costs were lower for larger systems (e.g., $3.9/W for systems >100 kW installed in 2008, compared to $4.5/W for systems 10-100 kW and $4.6/W for systems ≤10 kW).

Table 4. Module, Inverter, and Other Costs over Time

Installation Year	All System Sizes (capacity-weighted average)			≤ 10 kW			10-100 kW			>100 kW		
	Mod.	Inv.	Oth.	Mod.	Inv.	Oth.	Mod.	Inv.	Oth.	Mod.	Inv.	Oth.
2007	$4.3	$0.6	$2.8	$4.8	$0.7	$2.9	$4.7	$0.7	$2.7	$3.8	$0.4	$3.0
2008	$4.1	$0.5	$2.9	$4.6	$0.7	$2.9	$4.5	$0.6	$2.6	$3.9	$0.5	$3.0
2009	$4.0	$0.6	$3.0	$4.2	$0.8	$3.2	$4.2	$0.6	$2.7	$4.1	$0.5	$3.3

Notes: The results presented in this table are based on component level cost data provided by the California Solar Initiative, Minnesota's Solar Electric Rebate Program, and the Wisconsin Focus on Energy Cash-Back Rewards Program.

"Other" (non-module/non-inverter) costs vary with system size, and tend to be lowest for systems in the 10-100 kW size range. Among systems installed in 2009, for example, other costs were \$2.7/W for 10-100 kW systems, compared to \$3.2/W for ≤10 kW systems and \$3.3/W for >100 kW systems. This trend mirrors the trend shown in Figure 10, which demonstrated that economies of scale reduced average costs for systems up to 30 kW, but total installed costs increased somewhat for systems within the 30-250 kW size range.

Table 4 also illustrates recent temporal trends in component level costs, among the three programs that reported such information. Across all system sizes, the capacity-weighted average reported module cost declined by \$0.1/W from 2008 to 2009. However, when segmented by system size, average module costs declined only among ≤10 kW systems (by \$0.4/W) and 10-100 kW systems (by \$0.3/W), but increased by \$0.2/W among >100 kW systems. "Other" (nonmodule/non-inverter) costs increased from 2008-2009 across all size ranges (by \$0.1/W for all systems, by \$0.3/W for ≤10 kW systems and >100 kW systems, and by \$0.1/W for 10-100 kW systems).

Average Installed Costs for Residential Systems Are Lower in Germany and Japan than in the United States

Notwithstanding the significant cost reductions that have already occurred in the United States, international experience suggests that greater near-term cost reductions may be possible. Figure 13 compares average installed costs, excluding sales or value-added tax, in Germany, Japan, and the United States, focusing specifically on *small* residential systems (either 2-5 kW or 3-5 kW, depending on the country) installed in 2009. Among this class of systems, average installed costs were substantially lower in Germany and Japan (\$4.7/W and \$5.9/W, respectively) than in the United States (\$7.7/W). These differences may be partly attributable to the much greater cumulative grid-connected PV capacity in Germany and Japan (about 10,000 MW and 2,500 MW, respectively, at the end of 2009), compared to approximately 1,300 MW in the United States. That said, larger market size, alone, is unlikely to account for the entirety of the differences in average installed costs among countries.[27]

Installed Costs Vary Widely Across States

The United States is clearly not a homogenous PV market, as evidenced by Figure 14, which compares the average installed cost of systems ≤10 kW completed in 2009, across the 16 states in our dataset. Among systems in this size class, average costs range from a low of \$7.1/W in Texas to a high of \$9.6/W in Minnesota. Table 5 presents the same data in tabular form, along with comparative data for other system size ranges and groupings.

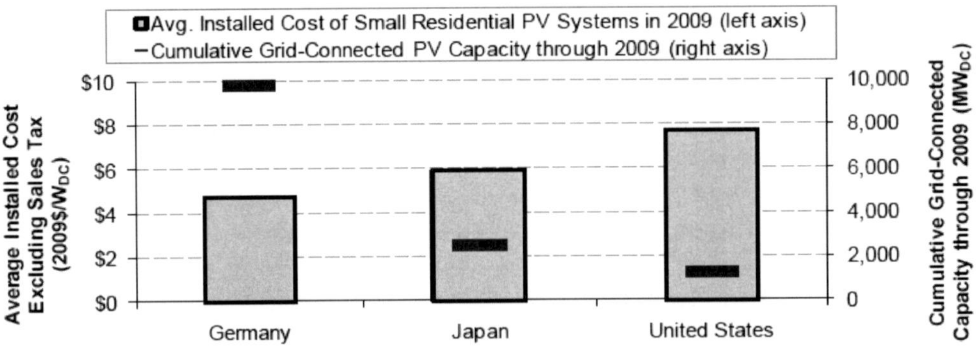

Notes: The Japanese and U.S. cost data are for 2-5 kW systems, while the German cost data are for 3-5 kW systems. Source for Japanese price and cumulative installed capacity data: Yamamoto, M. and O. Ikki. 2010. National Survey Report of PV Power Applications in Japan 2009. Paris, France: International Energy Agency Cooperative Programme on Photovoltaic Power Systems. Source for German price and cumulative installed capacity data: Wissing, L. 2010. National Survey Report of PV Power Applications in Germany 2009. Paris: France: International Energy Agency Cooperative Programme on Photovoltaic Power Systems.

Figure 13. Comparison of Average Installed Costs (Pre-Sales Tax/VAT) for Small Residential PV Systems in Germany, Japan, and the United States.

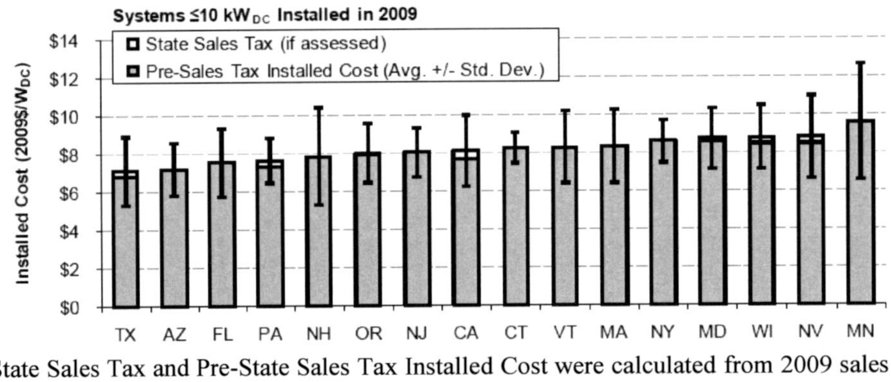

Notes: State Sales Tax and Pre-State Sales Tax Installed Cost were calculated from 2009 sales tax rates in each state (local sales taxes were not considered). Sales tax was assumed to have been assessed only on hardware costs, which, in turn, were assumed to constitute 65% of the total pre-sales-tax installed cost.

Figure 14. Variation in Installed Costs among U.S. States.

The variation in average installed costs across states may partially be a consequence of the differing size and maturity of the PV markets, where larger markets stimulate greater competition and hence greater efficiency in the delivery chain, and may also allow for bulk purchases and better access to lower-cost products. It is not surprising, therefore, that the two largest PV markets in the country (California and New Jersey) have relatively low average costs. This fact lends some credence to the premise behind state and local policies and programs that seek to reduce the cost of PV by accelerating deployment.

That said, the lowest cost states (Texas, Arizona, and Florida) are relatively small markets, illustrating the potential influence of other state- or local factors on installed costs.[28]

For example, administrative and regulatory compliance costs (e.g., incentive applications, permitting, and interconnection) can vary substantially across states, as can installation labor costs. Installed costs also vary across states as a result of differing sales tax treatment; eight of the 16 states shown in Figure 14 exempted residential PV systems from state sales tax throughout the entirety of 2009, and Oregon and New Hampshire have no state sales tax. Assuming that PV hardware costs represent approximately 65% of the total installed cost of residential PV systems (an assumption supported by the component-level cost data presented previously), state sales tax exemptions effectively reduce the post-sales-tax installed cost by $0.2-0.4/W, depending on the specific state sales tax rate that would otherwise be levied. Average installed costs may also differ among states due to differences in the proportion of systems that are ground-mounted or that have tracking equipment, both of which will tend to increase total installed cost.[29]Finally, as discussed further in Text Box 2, apparent differences in installed costs among states may also be an artifact of inconsistencies in how the installed cost of third-party owned systems is reported to PV incentive program administrators, although the available evidence does not suggest that this issue has significantly skewed the results.

Installed Costs Are Generally Lowest for Residential Systems and Highest for Public Sector Systems, Compared to Other Similarly Sized Systems

Figure 16 compares average installed costs across four customer segments: residential, commercial, non-profit, and public sector (i.e., government and schools). We focus on systems installed in 2009 for which customer segment data was provided, splitting those data into three size categories: 5-10 kW, 10-100 kW, and 100-500 kW.[30] The figure shows fairly consistent trends across all three system size ranges, in terms of the relative installed cost of PV systems among market segments. In general, installed costs tend to be lowest for residential systems compared to other similarly sized systems. Specifically, within the 5-10 kW size range, residential systems have an average installed cost ($7.6/W) that is $0.5/W less than commercial customers ($8. 1/W), $1 .2/W less than non-profit customers ($8. 8/W), and $1.1 /W less than public sector customers ($8.7/W). Consistent, though somewhat smaller, cost differentials are evident for systems in the 10-100 kW size range.

Figure 16 also indicates that public sector systems tend to have higher installed costs than similarly sized commercial systems. Specifically, the average installed cost of public sector systems exceeds that of commercial systems by $0. 6/W within the 5-10 kW system size range ($8.7/W vs. $8.1/W), by $0.7/W in the 10-100 kW size range ($8.6/W vs. $7.9/W), and by $0.8/W in the 100-500 kW size range ($8.1/W vs. $7.3/W). Non-profit systems also tend to have higher costs than similarly sized commercial systems, though the sample of non-profit systems in each size range is fairly limited.

Table 5. Average Installed Cost ($/WDC) by State and PV System Size Range

State	All Reported Yrs. Capacity-Weighted Average Cost (all sizes)		2009 Systems									
			Capacity-Weighted Average Cost (all sizes)		Simple Average Cost							
					0 - 10 kWDC		10 - 100 kWDC		100 - 500 kWDC		>500 kWDC	
AZ	$7.2	(n=3330)	$7.1	(n=2048)	$7.2	(n=1858)	$6.9	(n=187)	*	(n=3)	*	(n=0)
CA	$7.7	(n=58991)	$7.6	(n=15376)	$8.1	(n=13882)	$7.5	(n=1326)	$8.1	(n=106)	$7.2	(n=62)
CT	$7.9	(n=946)	$7.6	(n=306)	$8.3	(n=226)	$8.1	(n=61)	$7.3	(n=19)	*	(n=0)
FL	$7.5	(n=577)	$7.5	(n=575)	$7.6	(n=536)	$7.3	(n=38)	*	(n=0)	*	
MA	$8.1	(n=1990)	$7.4	(n=860)	$8.4	(n=740)	$8.0	(n=92)	$6.8	(n=26)	*	
MD	$9.0	(n=546)	$8.6	(n=316)	$8.8	(n=307)	$8.4	(n=9)	*	(n=0)	*	(n=0)
MN	$9.1	(n=198)	$9.3	(n=54)	$9.6	(n=49)	$9.6	(n=5)	*	(n=0)	*	(n=0)
NH	$7.6	(n=189)	$7.5	(n=157)	$7.9	(n=157)	*	(n=0)	*	(n=0)	*	(n=0)
NJ	$7.7	(n=4634)	$7.4	(n=1292)	$8.1	(n=964)	$7.9	(n=253)	$7.5	(n=62)	$7.2	(n=13)
NV	$8.7	(n=499)	$8.2	(n=183)	$8.8	(n=167)	$8.8	(n=16)	*	(n=0)	*	(n=0)
NY	$8.7	(n=1990)	$8.4	(n=779)	$8.6	(n=654)	$8.3	(n=125)	*	(n=0)	*	(n=0)
OR	$7.9	(n=1321)	$7.3	(n=473)	$8.0	(n=385)	$7.7	(n=76)	$6.9	(n=11)	*	(n=1)
PA	$7.9	(n=536)	$7.4	(n=372)	$7.7	(n=305)	$7.4	(n=66)	*		*	(n=0)
TX	$7.0	(n=1226)	$6.7	(n=459)	$7.1	(n=406)	$6.4	(n=51)	*		*	(n=0)
VT	$8.4	(n=365)	$7.9	(n=139)	$8.3	(n=134)	$7.2	(n=5)	*	(n=0)	*	(n=0)
WI	$8.7	(n=614)	$8.6	(n=264)	$8.8	(n=225)	$8.6	(n=39)	*	(n=0)	*	(n=0)

* Cost data are omitted if the sample size (n) is less than five.

Text Box 2. Reported Installed Costs for Customer-Owned vs. Third Party-Owned Systems

Third party ownership of customer-sited PV systems, through power purchase agreements and leases, has become increasingly common for PV systems of all sizes. Under such arrangements, the transaction between the host customer and the system owner typically consists of a series of payments over time, rather than a single up-front payment for the purchase of the PV system. The structure of the retail transaction therefore provides some discretion to third party PV system owners in terms of how they report installed cost to PV incentive program administrators. In some cases, the reported installed cost may include financing costs that are bundled into the payments made by the customer over time. Although customer- owned systems may also entail financing costs (e.g., if the customer finances some portion of the up-front payment through a home equity loan), those financing costs are incurred separate from the purchase of the PV system and therefore are not included in the installed cost data reported to PV incentive program administrators. Thus, to the extent that the installed cost of third party-owned systems reported to PV incentive program administrators includes financing costs or other costs unrelated to the installation of the PV system (e.g., maintenance or warranty services), it may obscure the underlying trends in PV installed costs.

To gain some insight into the potential significance of this issue, Figure 15 compares the reported installed cost of customer-owned and third party-owned systems in the California Solar Initiative (C SI) program. As shown, installed costs reported for third party-owned systems were, on average, higher than for customer-owned systems. Among all CSI systems installed in 2009, the capacity-weighted average installed cost reported for third party-owned systems was $0.4/W higher than for customer-owned systems. A similar price differential is evident for systems ≤10 kW, while for systems in the 10-100 kW size range, the average reported installed cost of third party-owned systems was $1 .2/W higher than for customer-owned systems. Figure 15 also identifies the average reported installed cost for all systems (both customer-owned and third party-owned). As shown, the higher reported cost for third party-owned systems – which represented about 13% of the CSI systems installed in 2009 – did not significantly affect the average cost data for all CSI systems. It therefore does not appear likely that any inconsistencies between how installed costs are reported for third party-owned and customer-owned systems significantly skewed the installed cost trends presented in this report. However, to the extent that third party ownership becomes a larger portion of the overall market, issues associated with consistent reporting of cost data could become more significant over time. It should also be noted that third party-owned systems may legitimately entail higher installed costs, if performance-enhancing equipment (e.g., tracking systems or advanced power electronics) or high-precision metering are more prevalent among those systems.

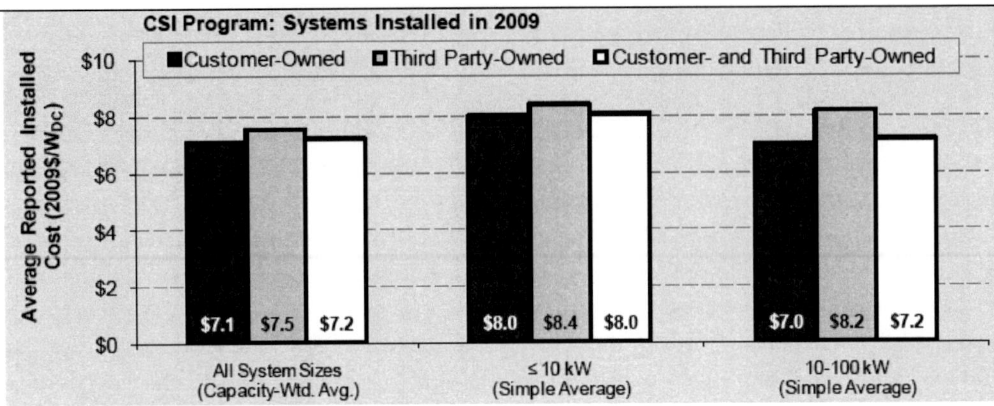

Figure 15. Comparison of Reported Installed Costs between Customer-Owned and Third Party-Owned Systems in the CSI Program.

The New Construction Market Offers Significant Cost Advantages for Residential PV, Despite the Higher Cost of BIPV Relative to Rack-Mounted Systems

Figure 17 compares the average installed cost of PV systems installed residential new construction and residential retrofit applications funded through two California programs (the CSI and New Solar Homes Partnership [NSHP] programs[31]), focusing in particular on 1-3 kW systems installed in 2009.[32] Among this group of PV systems, those installed in residential new construction cost $1 .6/W less, on average, than comparably-sized residential retrofit systems ($7.8/W compared to $9.4/W), a cost advantage of approximately 16%.

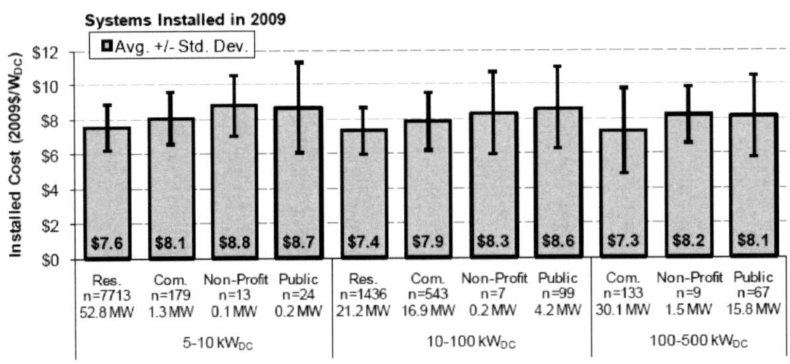

Figure 16. Variation in Installed Costs among Customer Sectors.

However, simply comparing the overall average cost of all residential new construction and all residential retrofit systems masks the fact that a much larger proportion of new construction systems are building-integrated PV (BIPV), which tend to have somewhat higher costs than rack-mounted systems, though the higher installed costs may be partially offset by avoided roofing material costs. Systems within the data sample were identified as BIPV or rack-mounted based on module manufacturer and model data provided by the PV incentive program administrators. As shown in Figure 17, BIPV systems in the two California programs cost $0.5/W more, on average, than rack- mounted systems installed in residential new construction (i.e., $8.0/W vs. $7.5/W).

To make an apples-to-apples comparison between residential new construction and residential retrofit applications, one can compare the average cost of rack-mounted systems installed in the two applications, as also shown in Figure 17. This comparison suggest a somewhat greater cost advantage for new construction than implied by the overall averages, with rack-mounted systems installed in residential new construction averaging $1. 9/W less than residential retrofit systems ($7.5/W compared to $9.4/W).

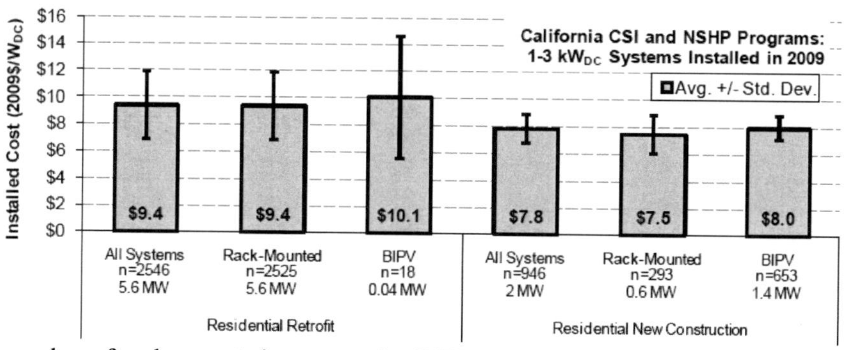

Note : The number of rack-mounted systems plus BIPV systems may not sum to the total number of systems, as some systems could not be identified as either rack-mounted or BIPV.

Figure 17. Comparison of Installed Costs for Residential Retrofit vs. New Construction.

Systems with Thin-Film Modules Had Higher Installed Costs than Those with Crystalline Modules

A number of countervailing factors may influence the relative cost of systems employing thinfilm[33] and crystalline modules. Thin-film modules are typically lower-cost than crystalline modules but are less efficient, which would tend to engender higher balance of system costs. In addition, greater uncertainty in the long-term performance of thin-film modules on the part of consumers and potentially faster degradation rates would tend to drive down the price of thin-film systems relative to crystalline systems. However, some thin-film technologies have higher energy yields (annual kWh per installed kW) than crystalline modules, due to better performance at high temperatures or under diffuse irradiance, which would tend to increase the price that customers are willing to pay for thin-film systems.

To understand the net effect of these cost drivers, Figure 18 compares the average installed cost of crystalline and thin-film systems, focusing specifically on rack-mounted (i.e., not BIPV) systems installed in 2009.[34] Individual systems within the dataset were identified as employing either thin- film or crystalline modules based on module manufacturer and model data provided by the PV incentive programs. As shown, thin-film systems in both of the size ranges shown had average installed costs higher than comparably-sized crystalline systems ($0.8/W higher among ≤10 kW systems and $0.4/W higher among 10-100 kW systems).[35]

To examine the relationship between total installed cost and module efficiency more directly, Figure 19 compares the average installed cost of all rack-mounted systems installed in 2009, according to module efficiency. Within both of the two system size ranges shown, average installed costs were lowest for systems with modules of mid-range efficiencies (14-16%), although the differences in installed cost across module efficiencies are relatively small. Among systems ≤10 kW, for example, average installed costs were lowest ($7.4/W) for systems with module efficiencies of 15-16%, compared to $8.2/W for systems with module efficiencies ≤12% and $8.3/W for systems with module efficiencies >18%.

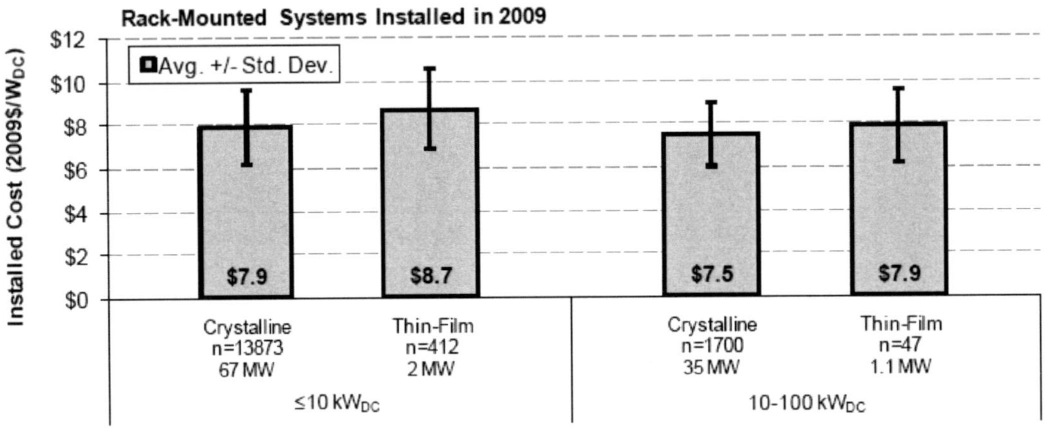

Figure 18. Comparison of Installed Costs for Crystalline vs. Thin-Film Systems.

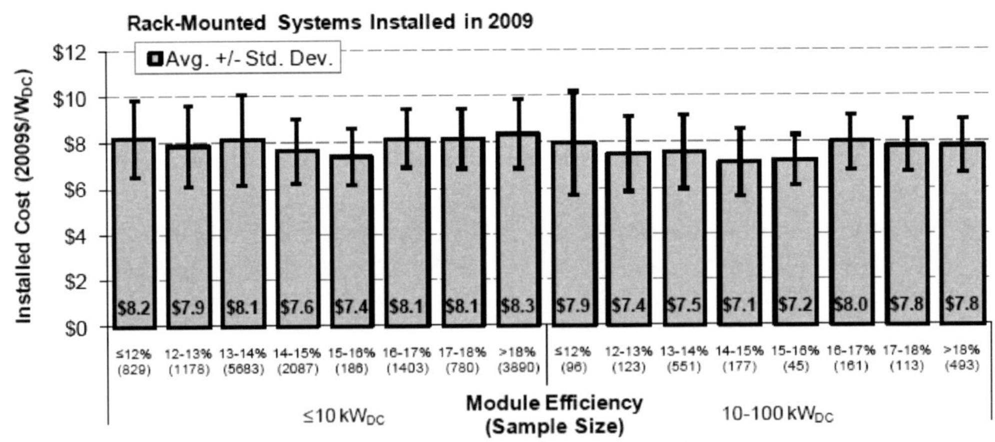

Figure 19. Variation in Total Installed Cost by Module Efficiency.

Tracking Systems Had Higher Installed Costs than Fixed-Axis Systems

Data indicating whether or not PV systems had tracking equipment were provided for a relatively small percentage of systems in the sample (14% of systems and 18% of capacity installed in 2009). Based on the limited data available, Figure 20 compares the average cost of PV systems with tracking (including both single- and double-axis tracking) to those with fixed mounting, focusing on rack-mounted systems (both roof- and ground-mounted) installed in 2009.[36] As shown, tracking systems had higher installed costs within both size categories, as would be expected, given the additional cost associated with the tracking equipment and ground-mounting.[37] Among systems ≤10 kW, tracking systems had average installed costs \$1.7/W (or 21%) higher than fixed-axis systems. This comports reasonably well with component-level cost data from Wisconsin's Focus on Energy Renewable Energy Cash-Back Rewards Program, which show that, among systems installed in 2009, those with tracking systems had "rack" costs \$1.3/W higher than fixed-axis systems.[38] Among the 10-100 kW systems summarized in Figure 20, tracking systems had average installed costs \$3.5/W (or 44%) higher than their fixed-axis counterparts; however, given the small number of tracking systems in this size range, caution is warranted in generalizing from these results.

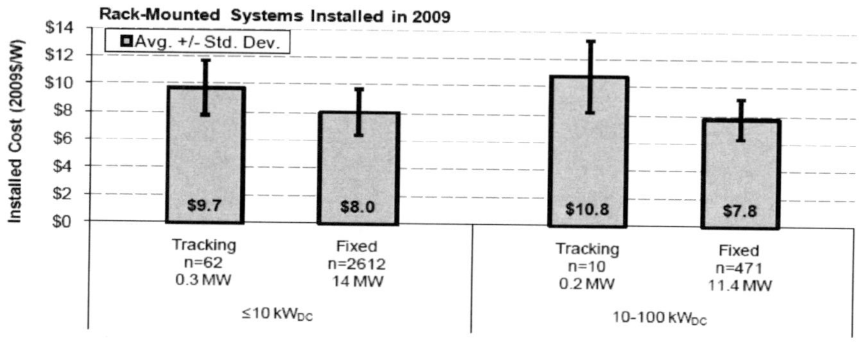

Figure 20. Comparison of Installed Costs for Tracking vs. Fixed-Axis Systems.

4. PV Incentive and Net Installed Cost Trends

Financial incentives provided through utility, state, and federal programs have been a major driving force for the PV market in the United States. For any individual system, these incentives potentially include some combination of cash incentives provided through state or utility PV incentive programs, federal and/or state investment tax credits (ITCs)[39], revenues from the sale of renewable energy certificates (RECs), and accelerated depreciation of capital investments in solar energy systems. This section describes trends in incentive levels (focusing specifically on state/utility cash incentives and state/federal ITCs) and net installed costs (i.e., installed costs after receipt of financial incentives) over time, by system size, and among states.

Two important caveats should be noted at the outset:

- First, the set of incentives addressed in this section are necessarily limited in scope, accounting only for the direct cash incentives provided through the specific state/utility PV incentive programs in the dataset, plus state and federal ITCs. The analysis does not account for the incentive for commercial PV provided through accelerated depreciation,[40] nor for any additional incentives that projects may have received from state/utility incentive programs outside of the PV incentive program covered in this report. The results presented in this section also do not account for revenue from the sale of RECs, although the potential magnitude of this revenue stream is briefly discussed in general terms (see Text Box 3). As such, the results presented in this section exclude New Jersey's SREC Registration Program (which is included in previous sections of this report), as that program provides incentives solely in the form of solar RECs, the price of which varies over time according to market conditions.

- Second, this section marks a departure from Section 3 by going beyond a simple reporting of data provided by program administrators. In particular, a variety of assumptions, as documented within this section and described further in Appendix C, were required in order to estimate the value of federal and state ITCs for each project and to determine the net installed cost on an after-tax basis.

State/Utility Cash Incentives Continued their Decline in 2009

The PV incentive programs represented within the dataset provide cash incentives of varying forms. Most provide up-front cash incentives (i.e., "rebates"), based either on system capacity, a percentage of installed cost, or a projection of annual energy production. Several programs, instead, provide performance-based incentives (PBIs), which are paid out over time based on actual energy production, as either a supplement or an alternative to an up-front rebate.[41] Figure 21 shows the average cash incentive received by the PV systems in the dataset, over time and according to system size. These data are presented on a *pre-tax* basis – that is, prior to assessment of state or federal taxes that may be levied if the incentive is treated as taxable income.[42] Note also that the figure does not necessarily provide an accurate depiction of the size of incentives *offered* in each year, as there is typically some lag between the time that a project reserves its incentives and the time that it is installed.

As shown in Figure 21, average cash incentives for systems installed in 2009 ranged from \$1.2/W - \$2.2/W across the system size categories shown, and were up to \$0.3/W lower than the incentive levels in the preceding year. Incentive levels in 2009 were 37-47% of their peak in 2002 and 2003, declining at an average rate of about \$0.3-0.4/W per year over the intervening period. These trends largely reflect changes in incentives received by systems funded by California's ERP, Self Generation Incentive Program (SGIP), and CSI program, and by New Jersey's Customer Onsite Renewable Energy (CORE) program, which together represent 79% of all of the systems in the data sample. To some extent, the trends in Figure 21 also reflect the increasing geographic diversification of the sample over time, with a progressively larger percentage of the sample consisting of systems funded by programs other than the three aforementioned California programs. In general, these other programs have tended to offer higher incentives, counteracting to some degree the declining incentives within the California programs. That said, average incentives among all PV incentive programs in the sample have generally declined over time (see Table B-3 in Appendix B).

Text Box 3. Revenue from the Sale of RECs

PV system owners may be able to sell RECs generated by their system, adding to any direct incentives received from state/utility PV incentive programs and Federal or state ITCs (provided that REC ownership is not automatically transferred to the state/utility as a condition of providing a direct cash incentive). Projecting the value of REC sales over the lifetime of each individual PV system in our dataset would be a highly speculative task, and therefore was not undertaken for this study. Based on recent REC prices, however, the revenue potential in most states is relatively modest, compared to the value of direct cash incentives received through state/utility PV incentive programs and to the value of the Federal ITC.

In general, the potential REC revenue for customer-sited PV depends on where the system is located, and consequently, what types of REC markets are available.

- Voluntary REC Markets. In most states, RECs generated by PV systems may be sold to individuals, businesses, or government agencies that are voluntarily seeking to support renewable energy. Given the elective nature of these transactions, prices in voluntary REC markets have historically been quite modest. For example, voluntary RECs traded through Spectron, a brokerage firm, averaged about $1.4/MWh in 2009. If extrapolated over 20 years, revenue from REC sales at this price would be equivalent to an up-front, pre-tax incentive of just $0.02/WDC on a present value basis (assuming a 10% nominal discount rate, an AC capacity factor of 20%, and DC-to-AC losses of 20%).
- General RPS Markets. In some states, RECs generated by PV systems may be sold to electricity suppliers for compliance with state renewables portfolio standards (RPS). These markets may offer greater REC revenue potential than in voluntary markets, though REC prices in RPS markets have historically varied quite substantially across states and over time. For PV, the most critical issue typically is whether the state RPS has a specific solar requirement (i.e., a solar "set-aside" or "carve- out"). In "general" RPS markets without a solar set-aside (in which case RECs from PV systems may be used to satisfy the total renewable electricity compliance obligation), the highest average REC prices in 2009 occurred in Massachusetts, where REC prices for compliance with the state's Class I RPS requirement averaged approximately $30/MWh (again, based on REC trades through Spectron). If extrapolated over a 20-year period, using the same assumptions as before, revenue from REC sales at this price would be equivalent to an up-front, pre-tax payment of $0.40/WDC.
- RPS Solar Set-Aside Markets. Substantially greater REC revenue potential may be available in RPS states with a solar (or distributed generation) set-aside. During 2009, the highest solar REC (or SREC) prices occurred for compliance with New Jersey's solar set-aside, with SREC prices averaging $542/MWh over the year. Extrapolating this revenue stream over a 15-year period (as PV systems in New Jersey can sell SRECs only for up to 15 years) yields the equivalent of an up-front, pre-tax payment of $6.4/W, which is significantly larger than the direct cash incentives offered in most states.

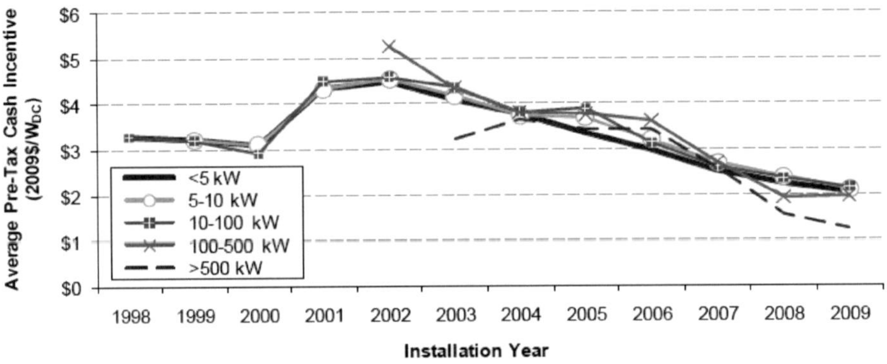

Notes: Averages shown only if five or more observations were available for a given size category in a given year. For sample sizes, refer to Table B-2 in the Appendix.

Figure 21. Pre-Tax State/Utility Cash Incentive Levels over Time.

In 2009, the Combined Value of Federal and State ITCs Plus Direct Cash Incentives Was Near Its Peak for Commercial PV, and up Significantly for Residential PV

Although direct cash incentives received from state and utility PV programs have, on average, declined over time, other sources of financial incentives have become more significant. Most notably, starting January 1, 2006, the federal ITC for commercial PV systems rose from 10% to 30% of project costs, and a 30% ITC (capped at $2,000) was established for residential PV. [43] The *Energy Improvement and Extension Act of 2008* subsequently lifted the $2,000 cap on the residential ITC for systems installed on or after January 1, 2009. In addition to the federal tax credits, a number of states have, at various times, offered state ITCs for PV, although these tax credits have generally been smaller and/or available to a more-restricted set of projects than the federal tax credit (see Appendix C for details on the ITCs for PV offered by the states in our dataset).

Figure 22 illustrates the combined effect of changes over time in state and federal ITCs (assuming that all customers take advantage of available tax credits) *plus* changes to the cash incentives provided through the state and utility PV incentive programs in the dataset, expressed here on an *after-tax* basis.[44] As noted previously, this assessment ignores potential revenues from the sale of RECs, though for most of the states in our dataset (other than New Jersey), such revenues would likely add only marginally to the overall incentive received (see Text Box 3).

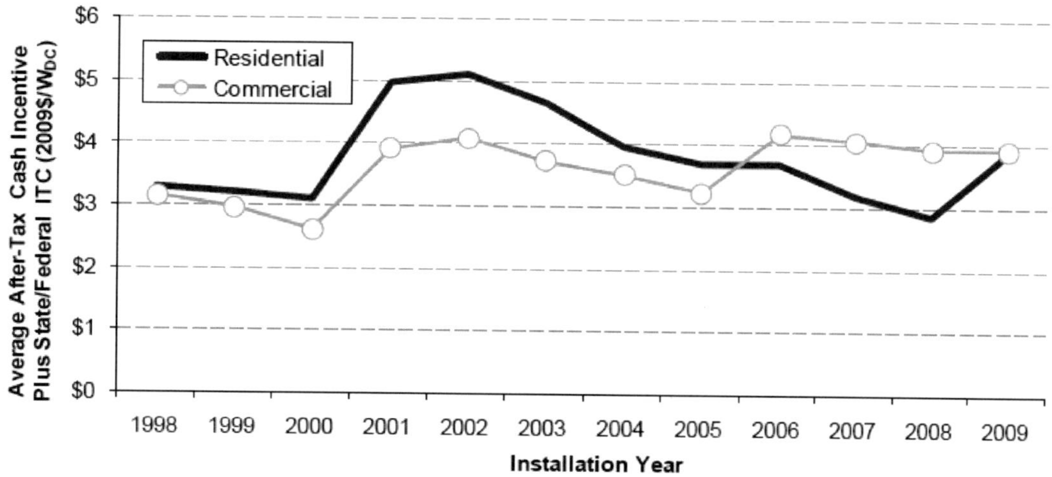

Notes: We assume that all systems ≤10 kW are residential and all systems >10 kW are commercial (unless indentified otherwise). For residential systems, we assume that state/utility cash incentives are non-taxable and reduce the basis of the federal ITC. For commercial systems, we assume that state/utility cash incentives are taxed at a federal corporate tax rate of 35% plus the prevailing state corporate tax rate, and do not reduce the basis of the federal ITC. The value of state ITCs is calculated as described in Appendix C.

Figure 22. After-Tax State/Utility Cash Incentives plus State & Federal ITCs (Calculated).

Figure 22 depicts a discernibly different trend than Figure 21, reflecting the introduction and subsequent increases in the federal ITC, occurring at different times for residential and commercial systems. Among residential PV systems, the decline in average aggregate incentive levels temporarily abated in 2006 when the federal ITC was extended to these systems, though the effect through 2008 was relatively small due to the $2,000 cap on the residential credit. After the cap on the residential ITC was lifted in 2009, the average aggregate incentive for residential PV rose substantially to $3.9/W, approximately a 37% increase over the average in 2008.

For commercial PV systems, the decline in the aggregate after-tax incentive that began in 2002 abruptly reversed course in 2006, when the federal ITC for commercial PV increased from 10% to 30% of project costs. The aggregate commercial incentive has subsequently declined slowly over time as cash incentives have been reduced; as of 2009, the average aggregate after-tax financial incentive received by commercial PV systems was $3.9/W, virtually unchanged from the prior year and slightly below its peak of $4.2/W in 2006.[45] The fact that, over the 2005-2008 timeframe, combined after-tax incentives rose substantially for commercial PV while declining for residential PV may partially explain the shift towards the commercial sector within the U.S. PV market over that period. With the lifting of the cap on the federal ITC for residential PV in 2009, however, some movement back towards the residential sector has begun to occur.[46]

Net Installed Costs Dropped Sharply in 2009 for Residential PV, While Remaining Flat for Commercial PV

In 2009, average *net installed costs* – that is, installed costs minus the combined after-tax value of state/utility cash incentives and state/federal ITCs – stood at $4. 1/W for residential PV and $4.0/W for commercial PV. For residential PV, the average net installed cost in 2009 represents an historical low, and was $1.3/W (or 24%) below the 2008 average, reversing the trend of the preceding several years during which average net installed costs for residential PV had been slowly rising. As discussed in Section 3, average pre-incentive installed costs remained relatively stable from 2005 to 2009, while average state/utility cash incentives were declining. The net effect of these two trends, as illustrated in Figure 23, is that the net installed cost of residential PV rose by approximately $0.7/W from 2006 to 2008. The dramatic reversal of this trend in 2009 is attributable to the lifting of the dollar cap on the federal ITC for residential PV installations beginning in that year. Given the expectation that state/utility cash incentives will continue to be ratcheted down, future reductions in the net installed cost of residential PV will require that reductions in pre-incentive installed costs keep pace with declining cash incentives.

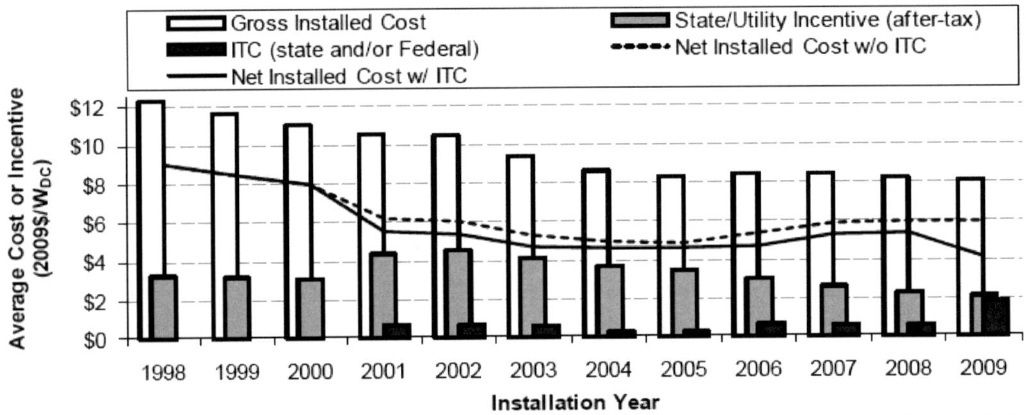

Notes: We assume that all systems ≤10 kW are residential , unless identified otherwise, and that state/utility cash incentives for residential PV are non-taxable and reduce the basis of the federal ITC. The value of state ITCs is calculated as described in Appendix C.

Figure 23. Net Installed Cost of Residential PV over Time (Calculated).

For commercial PV, the average net installed costs in 2009 was virtually unchanged from the preceding two years, and was up slightly (by $0.4/W or 11%) from the all-time low of $3 .7/W in 2006. The long-term trend for commercial PV differs from the trend for residential PV, by virtue of the fact that commercial PV systems had access to the full, uncapped federal ITC beginning in 2006. Like residential PV, however, future reductions in average net installed costs will require that reductions in pre-incentive installed costs keep pace with expected future reductions in state/utility cash incentives.

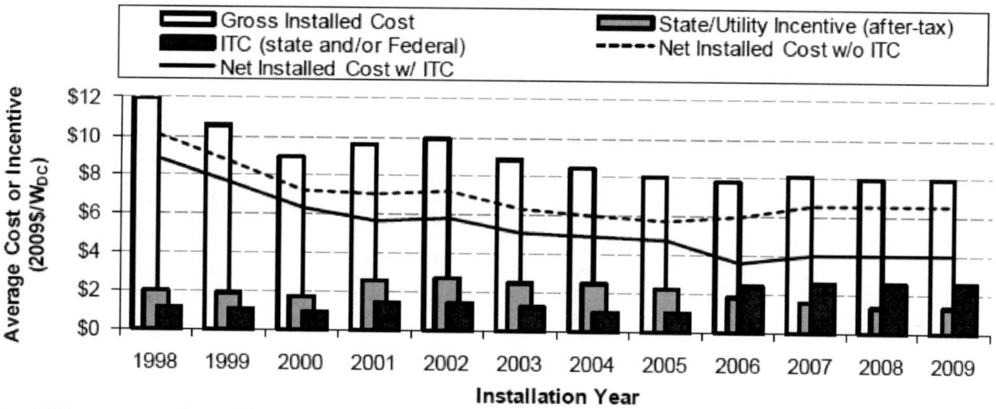

Notes: We assume that all systems >10 kW are commercial, unless identified otherwise, and that state/utility cash incentives for commercial PV are taxed at a federal corporate tax rate of 35% plus the prevailing state corporate tax rate, and do not reduce the basis of the federal ITC. The value of state ITCs is calculated as described in Appendix C.

Figure 24. Net Installed Cost of Commercial PV over Time (Calculated).

Finally, Figure 23 and Figure 24 also illustrate the potential impact of incentive levels on gross (i.e., pre-incentive) installed costs. A previous Berkeley Lab report found a statistically significant correlation can be seen in Figure 23 and Figure 24 (not surprisingly so, given the dominance of ERP and SGIP systems within the dataset). Most visibly, the decline in gross installed costs that had occurred during prior years ceased in 2001-2002, coinciding with a substantial increase in incentive levels under ERP and SGIP.

Incentives and Net Installed Costs Differ Widely across States

The preceding incentive-related trends are drawn from the entire dataset and are therefore dominated by the PV incentive programs in California and New Jersey. Incentives and net installed costs, however, vary significantly across all the states in the sample. Figure 25 and Figure 26 compare average incentive levels and net installed costs across states in 2009, for residential and commercial PV systems, respectively.[48] Again, note that this analysis does not capture all types of financial incentives that may be available to PV systems in each state (e.g., incentives offered by PV incentive programs outside of those included in the data sample, or revenue from the sale of RECs). In addition, systems participating in New Jersey's SREC Registration program are excluded from the analysis in this section, and the New Jersey results presented in Figure 25 and Figure 26 are based solely on data from the state's Customer Onsite Renewable Energy (CORE) program and its successor, the Renewable Energy Incentive Program (REIP).[49] New Jersey's position within this analysis – especially among commercial PV systems – could look substantially different if systems funded through the SREC Registration Program were included, and if the value of SRECs (which have significant value in New Jersey, as discussed in Text Box 3) were included.

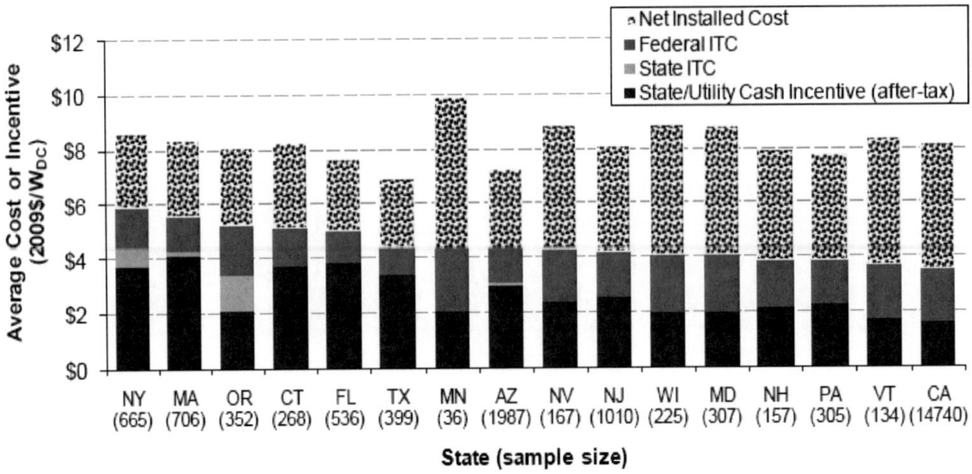

Notes: We assume that all systems ≤10 kW are residential unless identified otherwise, and that direct cash incentives for residential PV are non-taxable and reduce the basis of the federal ITC. State ITCs are calculated as described in Appendix C. Results shown for NJ are based solely on systems funded through the CORE and REIP programs.

Figure 25. Comparison of Incentive Levels and Net Installed Cost across States for Residential PV Systems Installed in 2009 (Calculated).

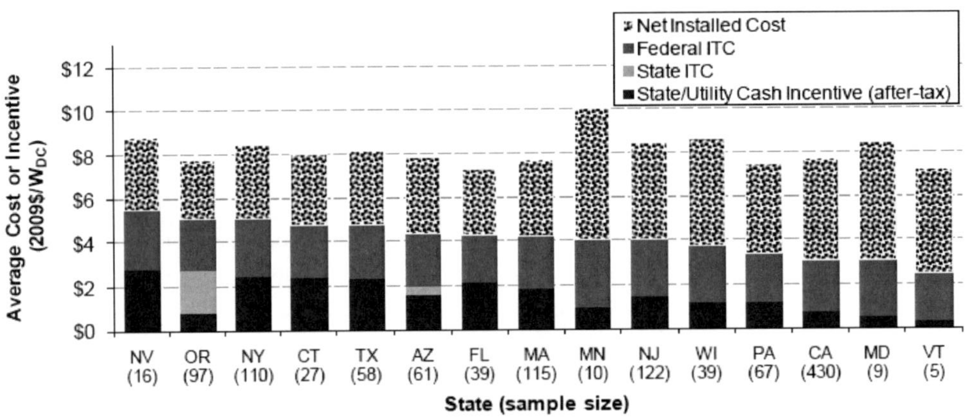

Notes: We assume that all systems >10 kW are commercial unless identified otherwise, and that direct cash incentives for commercial systems are taxed at a federal corporate tax rate of 35% plus the prevailing state corporate tax rate, and do not reduce the basis of the federal ITC. State ITCs are calculated as described in Appendix C. Results shown for NJ are based solely on systems funded through the CORE and REIP programs. NH is excluded from the figure, as the NH PV systems in the dataset are exclusively residential.

Figure 26. Comparison of Incentive Levels and Net Installed Cost across States for Commercial PV Systems Installed in 2009 (Calculated).

Among residential systems installed in 2009 (Figure 25), average after-tax incentives (i.e., direct cash incentives from state/utility PV incentive programs plus state and federal ITCs, but excluding revenue from sale of RECs) ranged from a low of $3.5/W in California to a high of $5.9/W in New York. Average net installed costs for residential PV were lowest in

Texas ($2.4/W), which had relatively lucrative incentives and the lowest average pre-incentive installed cost among the 16 states. At the other end of the spectrum was Minnesota, with an average net installed cost for residential PV equal to $5.5/W in 2009.

For commercial PV (Figure 26), average after-tax incentive levels and net installed costs also varied considerably across states in 2009.

Average after-tax incentives were lowest in Vermont ($2.5/W) and were highest in Nevada ($5.5/W). Average net installed costs for commercial PV in 2009 was lowest in Oregon ($2. 1/W), owing to its relatively lucrative incentives combined with relatively low pre-incentive installed costs, while Minnesota had the highest average net installed cost ($6.0/W), as a result of relatively low cash incentives and high pre-incentive installed costs.

CONCLUSION

The number of photovoltaic systems installed in the United States. has been growing at a rapid pace in recent years, driven in large measure by government incentives. Given the relatively high historical cost of PV, a key goal of these policies has been to encourage cost reductions over time. Out of this goal arises the need for reliable information on the historical installed cost of PV. To address this need, Lawrence Berkeley National Laboratory initiated a series of reports focused on describing trends in the installed cost of grid-connected PV systems in the United States. The present report, the third in the series, describes installed cost trends from 1998 through 2009, based on project-level data for approximately 78,000 grid-connected systems deployed across 16 states.

Available evidence confirms that the installed cost of customer-sited PV systems has declined substantially since 1998, though both the pace and source of those cost reductions have varied over time. Prior to 2005, installed cost reductions were associated primarily with a decline in non- module costs. Starting in 2005, however, cost reductions began to stall, as the supply-chain and delivery infrastructure struggled to keep pace with rapidly expanding global demand. Starting in 2008 and continuing into 2010, wholesale module prices began a steep downward trajectory, in response to expanded manufacturing capacity and the global financial crisis. These reductions in module prices, however, did not translate into a noticeable reduction in average installed costs for PV systems in 2009, perhaps reflecting a natural lag between the time that PV system installation contracts are signed and when systems are installed. Preliminary evidence does suggest, though, that average installed costs for PV systems installed in 2010 will be substantially lower than in 2009. Those trends will be more fully explored in the next edition of this report.

The historical trend towards declining installed costs, along with the narrowing of cost distributions, suggests that PV deployment policies have achieved some success in fostering competition within the industry and spurring improvements in the cost structure and efficiency of the PV delivery infrastructure. Moreover, the fact that states with the largest PV markets also appear to have somewhat lower average costs than most states with smaller markets lends credence to the premise that state and utility PV deployment policies can affect local costs. However, even lower average installed costs in Japan and Germany suggest that deeper near-term cost reductions may be possible. Indeed, further cost reductions will be necessary if the PV industry is to continue its expansion in the customer-sited market, given the desire of PV incentive programs to ratchet down the level of financial support offered to PV installations.

APPENDIX A: DATA CLEANING, CODING, AND STANDARDIZATION

To the extent possible, this report presents the data as provided directly by PV incentive program administrators; however, several steps were taken to clean the data and standardize it across programs, as described below.

Projects Removed from the Dataset: The initial data sample received from PV incentive program administrators consisted of 80,824 PV systems installed through 2009. To eliminate presumably erroneous numerical data entries, systems were removed from the dataset if the reported installed cost was less than $2/W (641 systems) or greater than $3 0/W (545 systems), or if the incentive amount was zero (57 systems) or greater than $30/W (236 systems). For the California Self Generation Incentive Program, systems receiving incentives from other subsidy programs were dropped (106 systems). In addition, systems missing installed cost data (1,545 systems), incentive data (929 systems), or system size data (278 systems) were removed from the dataset. Finally, 158 systems with battery back-up were removed from the dataset. In total, 2,872 systems from the initial sample were removed from the dataset as a result of these filters, yielding a final sample of 77,952 systems.

Manual Data Cleaning: Module manufacturer/model and inverter manufacturer/model data were reviewed in order to correct obvious misspellings and misidentifications, and to create standardized identifiers for individual module and inverter models.

Completion Date: The data provided by several PV incentive programs did not identify the system completion date. In lieu of this information, the best available proxy was used (e.g., the date of the incentive payment or the post-installation site inspection).

Identification of Residential New Construction and Residential Retrofit Systems: Section 3 compares the cost of systems installed in residential new construction to those installed in residential retrofit applications, focusing specifically on 1-3 kW systems installed through two California programs in 2009: the California Energy Commission (CEC)'s New Home Solar Partnership (NHSP) program and the California Solar Initiative (CSI). All systems installed through NHSP are assumed to be residential new construction, while all residential systems installed through CSI are assumed to be retrofit.

Identification of Building-Integrated and Rack-Mounted Residential Systems: The comparison between residential new construction and residential retrofit systems funded through NHSP and CSI is further differentiated between building-integrated PV (BIPV) and rack-mounted systems. The raw data provided by PV incentive program administrators did not include explicit identifiers for these categories; thus, systems were identified as either BIPV or rack-mounted by cross-referencing data provided on the module manufacturer and model for each system with the California Solar Initiative (CSI)'s List of Eligible Modules, which identifies whether modules are BIPV or rack-mounted.[50] Based on this procedure, 3,486 of the 3,492 applicable systems in the dataset (i.e., 1-3 kW systems funded through NHSP and CSI and installed in 2009) were identified as either BIPV or rack-mounted.

Identification of Crystalline and Thin-Film Systems: Section 3 compares the installed cost of systems with thin-film modules to those with crystalline modules. The raw data provided by PV program administrators generally do not include explicit identifiers for these categories. Thus, systems were categorized as crystalline, thin-film, or hybrid by cross-referencing data provided on module manufacturer and model with the CSI's List of Eligible Modules, which identifies whether modules are crystalline, thin- film, or hybrid. Based on this procedure, 21,497 of the 23,653 systems installed in 2009 were identified as employing either thin-film, crystalline, or hybrid modules.

Conversion to 2009 Real Dollars: Installed cost and incentive data are expressed throughout this report in real 2009 dollars (2009$). Data provided by PV program administrators in nominal dollars were converted to 2009$ using the "Monthly Consumer Price Index for All Urban Consumers," published by the U.S. Bureau of Labor Statistics.

Conversion of Capacity Data to Direct Current (DC) Watts at Standard Test Conditions (DC-STC): Throughout this report, all capacity and dollars-per-watt ($/W) data are expressed using DC-STC capacity ratings. Most programs directly provided data in units of DC-STC; however, four programs (the CEC's Emerging Renewables Program and the New Solar Home Partnership program, the CPUC's Self-Generation Incentive Program, and SMUD's Residential Retrofit and Commercial PV Programs) provided capacity data only in terms of the California Energy Commission Alternating Current (CEC-AC) rating convention. The CEC-AC rating represents peak AC power output at PVUSA Test Conditions (PTC). Capacity data from these four programs were converted to DC-STC, according to the procedures described below.

CEC Emerging Renewables Program (ERP), CEC New Solar Home Partnership (NSHP) Program, and SMUD Residential Retrofit and Commercial PV Programs: The data provided for these programs included data fields identifying the module manufacturer, model, and number of modules for most PV systems. DCSTC ratings were identified for most modules by cross-referencing the information provided about the module type with the CS I's List of Eligible Photovoltaic Modules, which identifies DC-STC ratings for most of the modules employed in the systems funded through these programs. The DC-STC rating for each module was then multiplied by the number of modules to determine the total DC-STC rating for the system, as a whole. This approach was used to determine the DC-STC capacity rating for all of the systems in the NSHP and SMUD datasets, and for 86% of the systems in the ERP dataset. For the remaining systems in the ERP dataset, either the module data fields were incomplete, or the module could not be cross-referenced with the CSI list, or the estimated DC-STC rating for the system was grossly inconsistent with the reported CECAC rating. In these cases, an average conversion factor of 1.200 $_{\text{WDC-STC/WCEC-AC}}$ was used, which was derived based on the averages for other systems in the ERP dataset.

CPUC Self-Generation Incentive Program (SGIP): The data provided for SGIP included data fields identifying module manufacturer and model (but not number of modules), and inverter manufacturer and model. DC-STC module ratings and DC-PTC module ratings (i.e., DC watts at PVUSA Test Conditions) were identified by cross-referencing the reported module type with the CSI' s List of Eligible Photovoltaic Modules. Similarly, the rated inverter efficiency for each project was identified by cross referencing the reported inverter type with the CSI's List of Eligible Inverters, which identifies inverter efficiency ratings for most of the inverters used within the systems funded through SGIP.[51] These pieces of information (module DC-STC rating, module DC-PTC rating, and inverter efficiency rating), along with the reported CEC-AC rating for the system, were used to estimate the system DC-STC rating according to the following:

SystemDC-STC = (SystemCEC-AC / Inverter Eff.) * (ModuleDC-STC / ModuleDC-PTC)

In cases where data on module manufacturer and model either was not provided or could not be matched with the CSI module list, then the DC-STC rating was calculated using the median ratio of module DC-STC to DC-PTC ratings for systems installed in the same year (0.88-0.90 WDC-STC/WDC-PTC). In cases where data on inverter manufacturer and model either was not provided or could not be matched with the CS I's inverter list, the inverter efficiency was stipulated based on the average inverter efficiency of systems in the SGIP dataset installed in the same year and for which inverter efficiency ratings could be identified. If neither the module nor inverter data were provided, then the DC-STC rating was calculated directly from the reported CEC-AC rating, using the median annual ratio of module DC-STC rating to system CEC-AC rating (1.19-1.22 WDC-STC/WCEC-AC).

Appendix B: Detailed Sample Size and Program-Level Summaries

Table B-1. Program-Level Annual Installation Data, Based on Final Study Sample

State	Program Administrator(s) and Program Name		1998	1999	2000	2001	2002	2003	2004	2005	2006	2007	2008	2009	Total
AZ	APS Solar & Renewables Incentive Program	*No. Systems*	-	-	-	-	4	10	42	73	183	231	369	1,379	*2,291*
		MW	-	-	-	-	0.0	0.1	0.2	0.4	1.1	1.4	3.0	8.9	*15.0*
	SRP EarthWise Solar Energy Program	*No. Systems*	-	-	-	-	-	-	-	27	124	107	112	669	*1,039*
		MW	-	-	-	-	-	-	-	0.1	0.5	0.6	1.6	4.8	*7.4*
CA	CEC Emerging Renewables Program (ERP)	*No. Systems*	39	178	213	1,238	2,246	2,964	4,540	3,862	6,117	5,862	688	-	*27,947*
		MW	0.2	0.7	0.9	4.8	9.8	15.1	22.4	20.4	34.2	34.3	3.6	-	*146.4*
	CEC New Solar Home Partnership (NSHP)	*No. Systems*	-	-	-	-	-	-	-	-	-	241	1,135	1,221	*2,597*
		MW	-	-	-	-	-	-	-	-	-	0.7	3.1	4.5	*8.3*
	CPUC California Solar Initiative (CSI)	*No. Systems*	-	-	-	-	-	-	-	-	-	3,549	8,297	13,228	*25,074*
		MW	-	-	-	-	-	-	-	-	-	33.3	136.7	156.0	*326.0*
	CPUC Self-Generation Incentive Program (SGIP)	*No. Systems*	-	-	-	-	16	71	147	191	143	141	87	10	*806*
		MW	-	-	-	-	2.3	11.6	17.3	26.9	29.5	33.4	24.6	2.9	*148.6*
	LADWP Solar Incentive Program	*No. Systems*	-	2	4	70	201	224	42	79	136	308	415	718	*2,199*
		MW	-	0.1	0.0	0.5	3.0	5.0	0.9	1.0	1.2	1.7	3.5	5.9	*22.9*
	SMUD Residential Retrofit and Commercial PV Programs	*No. Systems*	-	-	-	-	-	-	-	19	29	56	65	199	*368*
		MW	-	-	-	-	-	-	-	0.1	0.2	0.4	0.5	1.3	*2.5*
CT	CCEF Onsite Renewable DG Program	*No. Systems*	-	-	-	-	-	1	2	2	7	14	53	38	*117*
		MW	-	-	-	-	-	0.0	0.0	0.0	0.3	1.6	5.9	5.7	*13.6*
	CCEF Solar PV Program	*No. Systems*	-	-	-	-	-	-	-	32	89	168	272	268	*829*
		MW	-	-	-	-	-	-	-	0.1	0.4	0.9	1.7	1.8	*4.9*
FL	FEO Solar Energy System Incentives Program	*No. Systems*	-	-	-	-	-	-	-	-	1	-	1	575	*577*
		MW	-	-	-	-	-	-	-	-	0.0	-	0.0	4.6	*4.6*
MA	MassCEC Small Renewables Initiative*	*No. Systems*	-	-	-	-	1	70	127	91	258	214	369	860	*1,990*
		MW	-	-	-	-	0.0	0.3	0.6	0.8	1.8	1.6	2.8	12.4	*20.3*
MD	MEA Solar Energy Grant Program	*No. Systems*	-	-	-	-	-	-	-	7	43	45	135	316	*546*
		MW	-	-	-	-	-	-	-	0.0	0.2	0.1	0.5	1.4	*2.2*
MN	MSEO Solar Electric Rebate Program	*No. Systems*	-	-	-	-	1	8	23	12	24	39	37	54	*198*
		MW	-	-	-	-	0.0	0.0	0.1	0.0	0.1	0.2	0.1	0.3	*0.8*
NH	NHPUC Renewable Energy Rebate Program	*No. Systems*	-	-	-	-	-	-	-	-	-	-	32	157	*189*
		MW	-	-	-	-	-	-	-	-	-	-	0.1	0.5	*0.5*

Table B-1. (Continued)

State	Program Administrator(s) and Program Name		1998	1999	2000	2001	2002	2003	2004	2005	2006	2007	2008	2009	Total
NJ	NJCEP Customer Onsite Renewable Energy (CORE) Program	*No. Systems*	-	-	-	-	-	-	287	727	849	677	744	575	*3,859*
		MW	-	-	-	-	-	-	2.0	9.9	18.0	15.2	14.0	15.5	*74.5*
	NJCEP SREC Registration Program	*No. Systems*	-	-	-	-	-	-	-	-	-	2	56	76	*134*
		MW	-	-	-	-	-	-	-	-	-	0.0	8.4	18.5	*26.9*
	NJCEP Renewable Energy Incentive Program (REIP)	*No. Systems*	-	-	-	-	-	-	-	-	-	-	-	641	*641*
		MW	-	-	-	-	-	-	-	-	-	-	-	5.6	*5.6*
NV	NPC/SPPC RenewableGenerations Rebate Program	*No. Systems*	-	-	-	-	-	-	1	62	75	94	84	183	499
		MW	-	-	-	-	-	-	0.0	0.3	0.4	0.5	0.4	1.3	2.9
NY	NYSERDA PV Incentive Program	*No. Systems*	-	-	-	-	-	44	122	113	191	332	409	779	1,990
		MW	-	-	-	-	-	0.2	0.5	0.6	1.1	2.0	2.9	6.5	13.9
OR	ETO Solar Electric Program	*No. Systems*	-	-	-	-	-	56	134	85	130	206	237	473	1,321
		MW	-	-	-	-	-	0.3	0.6	0.3	0.6	1.0	3.5	6.0	12.3
PA	DEP Sunshine Solar PV Program (and other state agency programs**)	*No. Systems*	-	-	-	-	-	-	-	-	-	-	-	372	372
		MW	-	-	-	-	-	-	-	-	-	-	-	2.8	2.8
	SDF Solar PV Grant Program	*No. Systems*	-	-	-	-	3	17	28	23	54	21	18	-	164
		MW	-	-	-	-	0.0	0.1	0.1	0.1	0.2	0.1	0.1	-	0.7
TX	Austin Energy Power Saver Program	*No. Systems*	-	-	-	-	-	-	49	145	145	171	257	305	1,072
		MW	-	-	-	-	-	-	0.1	0.6	0.5	0.7	1.1	1.8	4.8
	IOU Solar Incentive Programs (AEP, Entergy, Oncor, SWEPCO, TNMP)	*No. Systems*	-	-	-	-	-	-	-	-	-	-	-	154	154
		MW	-	-	-	-	-	-	-	-	-	-	-	1.2	1.2
VT	RERC Small Scale Renewable Energy Incentive Program	*No. Systems*	-	-	-	-	-	-	33	15	24	61	93	139	365
		MW	-	-	-	-	-	-	0.1	0.0	0.1	0.2	0.4	0.5	1.3
WI	Focus on Energy Renewable Energy Cash-Back Rewards Program	*No. Systems*	-	-	-	-	6	9	12	22	62	96	143	264	614
		MW	-	-	-	-	0.0	0.0	0.0	0.1	0.2	0.5	0.8	1.6	3.3
	Total	*No. Systems*	*39*	*180*	*217*	*1,308*	*2,478*	*3,474*	*5,589*	*5,587*	*8,684*	*12,635*	*14,108*	*23,653*	*77,952*
		MW	*0.2*	*0.8*	*0.9*	*5.4*	*15.2*	*32.6*	*44.9*	*62.0*	*90.4*	*130.4*	*219.3*	*272.2*	*874.1*

* Included within the totals shown for MassCEC's Small Renewables Initiative are systems funded through predecessor PV incentive programs offered by the Massachusetts Technology Collaborative that targeted residential and small non-residential PV systems.

** Pennsylvania state agencies have offered various grant and rebate programs for renewable energy systems; for simplicity, these programs are summarized in aggregate. Of these programs, the largest funding source for PV systems through 2009 has been the Pennsylvania Department of Environmental Protection (DEP)'s Sunshine Program.

Table B-2. Sample Size by Installation Year and System Size Range

System Size Range	Installation Year												Total
	1998	1999	2000	2001	2002	2003	2004	2005	2006	2007	2008	2009	
No. Systems													
0-5 kW	31	156	180	1,108	1,877	2,258	3,490	3,176	4,982	7,200	8,139	12,280	44,877
5-10 kW	3	13	24	159	428	865	1,548	1,671	2,706	4,078	4,252	8,715	24,462
10-100 kW	5	10	12	36	152	308	508	637	882	1,206	1,363	2,349	7,468
100-500 kW	0	1	1	5	18	37	36	93	92	117	273	230	903
>500 kW	0	0	0	0	3	6	7	10	22	34	81	79	242
Total	*39*	*180*	*217*	*1,308*	*2,478*	*3,474*	*5,589*	*5,587*	*8,684*	*12,635*	*14,108*	*23,653*	*77,952*
Capacity (MW)													
0-5 kW	0.1	0.3	0.4	3.0	5.0	6.4	10.0	9.5	15.5	23.1	25.5	40.0	138.8
5-10 kW	0.0	0.1	0.2	1.0	2.8	5.8	10.5	11.8	18.7	28.3	29.1	59.2	167.5
10-100 kW	0.1	0.3	0.2	0.6	2.5	6.6	11.7	14.9	17.9	25.0	31.3	49.1	160.1
100-500 kW	0.0	0.1	0.1	0.8	3.2	8.5	7.5	18.4	20.6	27.0	65.9	51.5	203.6
>500 kW	0.0	0.0	0.0	0.0	1.7	5.3	5.1	7.4	17.7	27.1	67.5	72.3	204.2
Total	*0.2*	*0.8*	*0.9*	*5.4*	*15.2*	*32.6*	*44.9*	*62.0*	*90.4*	*130.4*	*219.3*	*272.2*	*874.1*

Table B-3. Annual Average Installed Cost and Direct Cash Incentives, by PV Incentive Program and System Size

State	Program Administrator and Program Name	Size Range		1998	1999	2000	2001	2002	2003	2004	2005	2006	2007	2008	2009
AZ	APS Solar and Renewables Incentive Program	≤10 kW	No. Systems	-	-	-	-	4	9	40	68	173	219	331	1266
			Avg. Cost	-	-	-	-	*	12.7	8.7	8.6	8.6	7.9	7.2	7.2
			Avg. Incentive	-	-	-	-	*	4.4	4.3	4.2	3.9	3.2	2.9	3.0
		10-100 kW	No. Systems	-	-	-	-	-	1	2	5	9	11	34	112
			Avg. Cost	-	-	-	-	-	*	*	11.3	8.4	9.0	6.8	7.0
			Avg. Incentive	-	-	-	-	-	*	*	3.6	4.1	3.6	2.8	2.8
		>100 kW	No. Systems	-	-	-	-	-	-	-	-	1	1	4	1
			Avg. Cost	-	-	-	-	-	-	-	-	*	*	*	*
			Avg. Incentive	-	-	-	-	-	-	-	-	*	*	*	*

Table B-3. (Continued)

State	Program Administrator and Program Name	Size Range		1998	1999	2000	2001	2002	2003	2004	2005	2006	2007	2008	2009
AZ	SRP EarthWise Solar Energy Program	≤10 kW	No. Systems	-	-	-	-	-	-	-	27	122	103	96	592
			Avg. Cost	-	-	-	-	-	-	-	7.6	8.4	7.4	7.1	7.2
			Avg. Incentive	-	-	-	-	-	-	-	3.4	3.2	3.1	3.0	2.9
		10-100 kW	No. Systems	-	-	-	-	-	-	-	-	2	4	14	75
			Avg. Cost	-	-	-	-	-	-	-	-	*	*	7.4	6.8
			Avg. Incentive	-	-	-	-	-	-	-	-	*	*	2.9	2.9
		>100 kW	No. Systems	-	-	-	-	-	-	-	-	-	-	2	2
			Avg. Cost	-	-	-	-	-	-	-	-	-	-	*	*
			Avg. Incentive	-	-	-	-	-	-	-	-	-	-	*	*
CA	CEC Emerging Renewables Program (ERP)	≤10 kW	No. Systems	34	168	200	1201	2107	2728	4184	3519	5498	5203	595	-
			Avg. Cost	12.3	11.6	11.0	10.5	10.5	9.3	8.5	8.2	8.3	8.4	8.1	-
			Avg. Incentive	3.3	3.2	3.1	4.3	4.3	3.9	3.5	2.9	2.6	2.4	2.3	-
		10-100 kW	No. Systems	5	9	12	33	135	234	356	343	619	659	93	-
			Avg. Cost	12.0	11.2	9.0	10.1	10.0	8.7	8.0	7.6	7.7	8.0	7.8	-
			Avg. Incentive	3.3	3.2	2.9	4.3	4.4	4.0	3.5	2.9	2.6	2.4	2.3	-
		>100 kW	No. Systems	-	1	1	4	4	2	-	-	-	-	-	-
			Avg. Cost	-	*	*	*	*	*	-	-	-	-	-	-
			Avg. Incentive	-	*	*	*	*	*	-	-	-	-	-	-
CA	CEC New Solar Home Partnership (NSHP)	≤10 kW	No. Systems	-	-	-	-	-	-	-	-	-	239	1120	1166
			Avg. Cost	-	-	-	-	-	-	-	-	-	8.1	8.0	7.8
			Avg. Incentive	-	-	-	-	-	-	-	-	-	2.3	2.3	2.4
		10-100 kW	No. Systems	-	-	-	-	-	-	-	-	-	2	15	53
			Avg. Cost	-	-	-	-	-	-	-	-	-	*	6.7	7.3
			Avg. Incentive	-	-	-	-	-	-	-	-	-	*	2.3	2.4
		>100 kW	No. Systems	-	-	-	-	-	-	-	-	-	-	-	2
			Avg. Cost	-	-	-	-	-	-	-	-	-	-	-	*
			Avg. Incentive	-	-	-	-	-	-	-	-	-	-	-	*
CA	CPUC California Solar Initiative (CSI)	≤10 kW	No. Systems	-	-	-	-	-	-	-	-	-	3253	7301	11872
			Avg. Cost	-	-	-	-	-	-	-	-	-	8.4	8.2	8.1
			Avg. Incentive	-	-	-	-	-	-	-	-	-	2.0	1.7	1.4

State	Program Administrator and Program Name	Size Range		1998	1999	2000	2001	2002	2003	2004	2005	2006	2007	2008	2009
		10-100 kW	No. Systems	-	-	-	-	-	-	-	-	-	267	781	1202
			Avg. Cost	-	-	-	-	-	-	-	-	-	8.1	7.7	7.5
			Avg. Incentive	-	-	-	-	-	-	-	-	-	2.0	1.7	1.4
		>100 kW	No. Systems	-	-	-	-	-	-	-	-	-	29	215	154
			Avg. Cost	-	-	-	-	-	-	-	-	-	7.1	7.4	7.8
			Avg. Incentive	-	-	-	-	-	-	-	-	-	1.6	1.3	1.1
CA	CPUC Self-Generation Incentive Program (SGIP)	≤10 kW	No. Systems	-	-	-	-	-	-	-	-	-	-	-	-
			Avg. Cost	-	-	-	-	-	-	-	-	-	-	-	-
			Avg. Incentive	-	-	-	-	-	-	-	-	-	-	-	-
		10-100 kW	No. Systems	-	-	-	-	10	44	108	107	72	53	31	4
			Avg. Cost	-	-	-	-	9.8	8.1	8.4	7.9	7.7	7.5	7.1	*
			Avg. Incentive	-	-	-	-	4.4	3.8	4.0	3.8	3.3	2.6	2.4	*
		>100 kW	No. Systems	-	-	-	-	6	27	39	84	71	88	56	6
			Avg. Cost	-	-	-	-	8.0	7.0	7.9	7.5	7.5	7.2	7.4	9.5
			Avg. Incentive	-	-	-	-	4.0	3.3	3.8	3.7	3.4	2.7	2.4	2.2
CA	LADWP Solar Incentive Program	≤10 kW	No. Systems	-	1	4	66	184	194	37	70	126	276	385	655
			Avg. Cost	-	*	*	11.0	10.8	9.6	8.8	8.0	8.7	8.8	8.4	8.4
			Avg. Incentive	-	*	*	5.7	6.4	5.9	3.7	3.2	3.5	3.7	3.6	3.5
		10-100 kW	No. Systems	-	1	-	3	6	17	3	6	7	32	25	60
			Avg. Cost	-	*	-	*	9.5	9.8	*	7.2	7.9	8.5	8.0	8.0
			Avg. Incentive	-	*	-	*	6.1	6.2	*	3.1	3.1	3.5	3.5	3.3
		>100 kW	No. Systems	-	-	-	1	11	13	2	3	3	-	5	3
			Avg. Cost	-	-	-	*	9.5	10.3	*	*	*	-	6.9	*
			Avg. Incentive	-	-	-	*	6.3	6.2	*	*	*	-	3.0	*
CA	SMUD Residential Retrofit and Commercial PV Programs	≤10 kW	No. Systems	-	-	-	-	-	-	-	16	27	53	63	189
			Avg. Cost	-	-	-	-	-	-	-	8.6	8.6	8.3	8.0	8.5
			Avg. Incentive	-	-	-	-	-	-	-	2.8	2.4	2.1	2.0	1.8
		10-100 kW	No. Systems	-	-	-	-	-	-	-	3	2	2	-	7
			Avg. Cost	-	-	-	-	-	-	-	*	*	*	-	8.0
			Avg. Incentive	-	-	-	-	-	-	-	*	*	*	-	1.9
		>100 kW	No. Systems	-	-	-	-	-	-	-	-	-	1	2	3
			Avg. Cost	-	-	-	-	-	-	-	-	-	*	*	*
			Avg. Incentive	-	-	-	-	-	-	-	-	-	*	*	*

Table B-3. (Continued)

State	Program Administrator and Program Name	Size Range		1998	1999	2000	2001	2002	2003	2004	2005	2006	2007	2008	2009
CT	CCEF Onsite Renewable DG Program	≤10 kW	No. Systems	-	-	-	-	-	1	1	1	1	2	10	2
			Avg. Cost	-	-	-	-	-	*	*	*	*	*	8.4	*
			Avg. Incentive	-	-	-	-	-	*	*	*	*	*	5.3	*
		10-100 kW	No. Systems	-	-	-	-	-	-	1	1	5	8	23	17
			Avg. Cost	-	-	-	-	-	-	*	*	8.8	8.5	8.2	8.5
			Avg. Incentive	-	-	-	-	-	-	*	*	4.8	4.6	4.3	4.1
		>100 kW	No. Systems	-	-	-	-	-	-	-	-	1	4	20	19
			Avg. Cost	-	-	-	-	-	-	-	-	*	*	7.7	7.3
			Avg. Incentive	-	-	-	-	-	-	-	-	*	*	4.2	3.8
CT	CCEF Solar PV Program	≤10 kW	No. Systems	-	-	-	-	-	-	-	32	88	162	243	224
			Avg. Cost	-	-	-	-	-	-	-	9.0	9.2	9.2	8.7	8.3
			Avg. Incentive	-	-	-	-	-	-	-	5.0	4.7	4.4	4.2	3.8
		10-100 kW	No. Systems	-	-	-	-	-	-	-	-	1	6	29	44
			Avg. Cost	-	-	-	-	-	-	-	-	*	8.5	8.3	8.0
			Avg. Incentive	-	-	-	-	-	-	-	-	*	3.8	4.1	3.6
		>100 kW	No. Systems	-	-	-	-	-	-	-	-	-	-	-	-
			Avg. Cost	-	-	-	-	-	-	-	-	-	-	-	-
			Avg. Incentive	-	-	-	-	-	-	-	-	-	-	-	-
FL	FEO Solar Energy System Incentives Program	≤10 kW	No. Systems	-	-	-	-	-	-	-	-	1	-	1	536
			Avg. Cost	-	-	-	-	-	-	-	-	*	-	*	7.6
			Avg. Incentive	-	-	-	-	-	-	-	-	*	-	*	3.9
		10-100 kW	No. Systems	-	-	-	-	-	-	-	-	-	-	-	38
			Avg. Cost	-	-	-	-	-	-	-	-	-	-	-	7.3
			Avg. Incentive	-	-	-	-	-	-	-	-	-	-	-	3.5
		>100 kW	No. Systems	-	-	-	-	-	-	-	-	-	-	-	1
			Avg. Cost	-	-	-	-	-	-	-	-	-	-	-	*
			Avg. Incentive	-	-	-	-	-	-	-	-	-	-	-	*
MA	MassCEC Small Renewables	≤10 kW	No. Systems	-	-	-	-	-	65	118	74	241	200	327	740
			Avg. Cost	-	-	-	-	-	10.5	9.2	9.3	9.5	9.3	8.6	8.4
			Avg. Incentive	-	-	-	-	-	4.9	5.0	5.0	4.2	4.2	3.8	4.0

State	Program Administrator and Program Name	Size Range		1998	1999	2000	2001	2002	2003	2004	2005	2006	2007	2008	2009
	Initiative	10-100 kW	No. Systems	-	-	-	-	1	5	9	17	14	12	38	92
			Avg. Cost	-	-	-	-	*	13.0	10.9	10.3	10.5	9.8	8.6	8.0
			Avg. Incentive	-	-	-	-	*	15.2	8.5	11.1	8.3	7.3	4.1	4.3
		>100 kW	No. Systems	-	-	-	-	-	-	-	-	3	2	4	28
			Avg. Cost	-	-	-	-	-	-	-	-	*	*	*	6.8
			Avg. Incentive	-	-	-	-	-	-	-	-	*	*	*	3.4
MD	MEA Solar Energy Grant Program	≤10 kW	No. Systems	-	-	-	-	-	-	-	7	42	44	131	307
			Avg. Cost	-	-	-	-	-	-	-	11.4	11.7	10.7	9.3	8.8
			Avg. Incentive	-	-	-	-	-	-	-	1.3	1.6	1.5	1.8	2.0
		10-100 kW	No. Systems	-	-	-	-	-	-	-	-	1	1	4	9
			Avg. Cost	-	-	-	-	-	-	-	-	*	*	*	8.4
			Avg. Incentive	-	-	-	-	-	-	-	-	*	*	*	0.9
		>100 kW	No. Systems	-	-	-	-	-	-	-	-	-	-	-	-
			Avg. Cost	-	-	-	-	-	-	-	-	-	-	-	-
			Avg. Incentive	-	-	-	-	-	-	-	-	-	-	-	-
MN	MSEO Solar Electric Rebate Program	≤10 kW	No. Systems	-	-	-	-	1	8	23	12	24	37	36	49
			Avg. Cost	-	-	-	-	*	9.3	7.8	9.5	8.6	9.4	9.9	9.6
			Avg. Incentive	-	-	-	-	*	2.3	2.2	2.2	2.1	2.1	2.0	2.1
		10-100 kW	No. Systems	-	-	-	-	-	-	-	-	-	2	1	5
			Avg. Cost	-	-	-	-	-	-	-	-	-	*	*	9.6
			Avg. Incentive	-	-	-	-	-	-	-	-	-	*	*	2.1
		>100 kW	No. Systems	-	-	-	-	-	-	-	-	-	-	-	-
			Avg. Cost	-	-	-	-	-	-	-	-	-	-	-	-
			Avg. Incentive	-	-	-	-	-	-	-	-	-	-	-	-
NH	NHPUC Renewable Energy Rebate Program	≤10 kW	No. Systems	-	-	-	-	-	-	-	-	-	-	32	157
			Avg. Cost	-	-	-	-	-	-	-	-	-	-	8.7	7.9
			Avg. Incentive	-	-	-	-	-	-	-	-	-	-	2.4	2.1
		10-100 kW	No. Systems	-	-	-	-	-	-	-	-	-	-	-	-
			Avg. Cost	-	-	-	-	-	-	-	-	-	-	-	-
			Avg. Incentive	-	-	-	-	-	-	-	-	-	-	-	-
		>100 kW	No. Systems	-	-	-	-	-	-	-	-	-	-	-	-
			Avg. Cost	-	-	-	-	-	-	-	-	-	-	-	-
			Avg. Incentive	-	-	-	-	-	-	-	-	-	-	-	-

Table B-3. (Continued)

State	Program Administrator and Program Name	Size Range		1998	1999	2000	2001	2002	2003	2004	2005	2006	2007	2008	2009
NJ	NJCEP Customer Onsite Renewable Energy (CORE) Program	≤10 kW	No. Systems	-	-	-	-	-	-	267	592	706	570	596	398
			Avg. Cost	-	-	-	-	-	-	9.2	8.7	8.6	8.6	8.4	8.4
			Avg. Incentive	-	-	-	-	-	-	6.2	5.9	5.4	4.8	4.2	3.7
		10-100 kW	No. Systems	-	-	-	-	-	-	19	119	108	81	130	148
			Avg. Cost	-	-	-	-	-	-	9.5	8.5	8.5	8.9	8.3	8.1
			Avg. Incentive	-	-	-	-	-	-	5.4	5.6	5.1	4.5	4.0	3.4
		>100 kW	No. Systems	-	-	-	-	-	-	1	16	35	26	18	29
			Avg. Cost	-	-	-	-	-	-	*	7.5	7.4	7.3	8.1	8.2
			Avg. Incentive	-	-	-	-	-	-	*	4.3	4.0	3.5	3.2	2.6
NJ	NJCEP SREC Registration Program**	≤10 kW	No. Systems	-	-	-	-	-	-	-	-	-	2	21	11
			Avg. Cost	-	-	-	-	-	-	-	-	-	*	8.0	7.9
			Avg. Incentive	-	-	-	-	-	-	-	-	-	-	-	-
		10-100 kW	No. Systems	-	-	-	-	-	-	-	-	-	-	15	19
			Avg. Cost	-	-	-	-	-	-	-	-	-	-	7.5	7.8
			Avg. Incentive	-	-	-	-	-	-	-	-	-	-	-	-
		>100 kW	No. Systems	-	-	-	-	-	-	-	-	-	-	20	46
			Avg. Cost	-	-	-	-	-	-	-	-	-	-	6.9	6.9
			Avg. Incentive	-	-	-	-	-	-	-	-	-	-	-	-
NJ	NJCEP Renewable Energy Incentive Program (REIP)	≤10 kW	No. Systems	-	-	-	-	-	-	-	-	-	-	-	555
			Avg. Cost	-	-	-	-	-	-	-	-	-	-	-	7.8
			Avg. Incentive	-	-	-	-	-	-	-	-	-	-	-	1.7
		10-100 kW	No. Systems	-	-	-	-	-	-	-	-	-	-	-	86
			Avg. Cost	-	-	-	-	-	-	-	-	-	-	-	7.6
			Avg. Incentive	-	-	-	-	-	-	-	-	-	-	-	1.3
		>100 kW	No. Systems	-	-	-	-	-	-	-	-	-	-	-	-
			Avg. Cost	-	-	-	-	-	-	-	-	-	-	-	-
			Avg. Incentive	-	-	-	-	-	-	-	-	-	-	-	-
NV	NPC/SPPC RenewableGenerations Rebate	≤10 kW	No. Systems	-	-	-	-	-	-	1	55	69	89	80	167
			Avg. Cost	-	-	-	-	-	-	*	9.4	9.0	9.6	9.2	8.8
			Avg. Incentive	-	-	-	-	-	-	*	5.1	3.8	3.1	2.8	2.4

State	Program Administrator and Program Name	Size Range		1998	1999	2000	2001	2002	2003	2004	2005	2006	2007	2008	2009
	Program	10-100 kW	No. Systems	-	-	-	-	-	-	-	7	6	5	4	16
			Avg. Cost	-	-	-	-	-	-	-	13.6	9.4	8.1	*	8.8
			Avg. Incentive	-	-	-	-	-	-	-	5.3	4.2	3.4	*	4.4
		>100 kW	No. Systems	-	-	-	-	-	-	-	-	-	-	-	-
			Avg. Cost	-	-	-	-	-	-	-	-	-	-	-	-
			Avg. Incentive	-	-	-	-	-	-	-	-	-	-	-	-
NY	NYSERDA PV Incentive Program	≤10 kW	No. Systems	-	-	-	-	-	38	113	97	170	306	362	654
			Avg. Cost	-	-	-	-	-	9.7	10.5	9.8	9.2	9.1	8.8	8.6
			Avg. Incentive	-	-	-	-	-	5.0	6.4	5.7	4.3	4.2	4.0	3.8
		10-100 kW	No. Systems	-	-	-	-	-	6	9	16	21	26	47	125
			Avg. Cost	-	-	-	-	-	9.5	8.5	8.3	9.3	9.3	8.8	8.3
			Avg. Incentive	-	-	-	-	-	5.6	5.4	4.5	4.4	4.1	4.1	3.9
		>100 kW	No. Systems	-	-	-	-	-	-	-	-	-	-	-	-
			Avg. Cost	-	-	-	-	-	-	-	-	-	-	-	-
			Avg. Incentive	-	-	-	-	-	-	-	-	-	-	-	-
OR	ETO Solar Electric Program	≤10 kW	No. Systems	-	-	-	-	-	54	132	82	123	191	193	385
			Avg. Cost	-	-	-	-	-	8.1	7.4	7.8	8.6	8.9	8.7	8.0
			Avg. Incentive	-	-	-	-	-	4.7	4.1	3.2	2.1	2.0	1.9	2.1
		10-100 kW	No. Systems	-	-	-	-	-	1	1	3	7	15	36	76
			Avg. Cost	-	-	-	-	-	*	*	*	7.4	9.2	9.5	7.7
			Avg. Incentive	-	-	-	-	-	*	*	*	1.2	1.5	1.3	1.4
		>100 kW	No. Systems	-	-	-	-	-	1	1	-	-	-	8	12
			Avg. Cost	-	-	-	-	-	*	*	-	-	-	8.2	6.8
			Avg. Incentive	-	-	-	-	-	*	*	-	-	-	1.3	1.0
PA	DEP Sunshine Solar PV Program (and other state agency programs)	≤10 kW	No. Systems	-	-	-	-	-	-	-	-	-	-	-	305
			Avg. Cost	-	-	-	-	-	-	-	-	-	-	-	7.7
			Avg. Incentive	-	-	-	-	-	-	-	-	-	-	-	2.2
		10-100 kW	No. Systems	-	-	-	-	-	-	-	-	-	-	-	66
			Avg. Cost	-	-	-	-	-	-	-	-	-	-	-	7.4
			Avg. Incentive	-	-	-	-	-	-	-	-	-	-	-	2.0
		>100 kW	No. Systems	-	-	-	-	-	-	-	-	-	-	-	1
			Avg. Cost	-	-	-	-	-	-	-	-	-	-	-	*
			Avg. Incentive	-	-	-	-	-	-	-	-	-	-	-	*

Table B-3. (Continued)

State	Program Administrator and Program Name	Size Range		1998	1999	2000	2001	2002	2003	2004	2005	2006	2007	2008	2009
PA	SDF Solar PV Grant Program	≤10 kW	No. Systems	-	-	-	-	3	17	28	23	53	21	16	-
			Avg. Cost	-	-	-	-	*	10.7	12.3	10.4	9.4	9.5	9.8	-
			Avg. Incentive	-	-	-	-	*	6.6	5.8	5.6	5.2	4.9	3.9	-
		10-100 kW	No. Systems	-	-	-	-	-	-	-	-	1	-	2	-
			Avg. Cost	-	-	-	-	-	-	-	-	*	-	*	-
			Avg. Incentive	-	-	-	-	-	-	-	-	*	-	*	-
		>100 kW	No. Systems	-	-	-	-	-	-	-	-	-	-	-	-
			Avg. Cost	-	-	-	-	-	-	-	-	-	-	-	-
			Avg. Incentive	-	-	-	-	-	-	-	-	-	-	-	-
TX	Austin Energy Power Saver Program	≤10 kW	No. Systems	-	-	-	-	-	-	49	135	141	162	240	270
			Avg. Cost	-	-	-	-	-	-	7.4	7.2	7.2	7.3	7.6	6.8
			Avg. Incentive	-	-	-	-	-	-	5.2	5.1	4.3	4.3	4.3	3.9
		10-100 kW	No. Systems	-	-	-	-	-	-	-	10	4	9	17	35
			Avg. Cost	-	-	-	-	-	-	-	6.8	*	9.1	7.6	6.4
			Avg. Incentive	-	-	-	-	-	-	-	5.2	*	4.5	4.0	4.0
		>100 kW	No. Systems	-	-	-	-	-	-	-	-	-	-	-	-
			Avg. Cost	-	-	-	-	-	-	-	-	-	-	-	-
			Avg. Incentive	-	-	-	-	-	-	-	-	-	-	-	-
TX	IOU Solar Incentive Programs (AEP, Entergy, Oncor, SWEPCO, TNMP)	≤10 kW	No. Systems	-	-	-	-	-	-	-	-	-	-	-	136
			Avg. Cost	-	-	-	-	-	-	-	-	-	-	-	7.8
			Avg. Incentive	-	-	-	-	-	-	-	-	-	-	-	2.4
		10-100 kW	No. Systems	-	-	-	-	-	-	-	-	-	-	-	16
			Avg. Cost	-	-	-	-	-	-	-	-	-	-	-	6.5
			Avg. Incentive	-	-	-	-	-	-	-	-	-	-	-	2.3
		>100 kW	No. Systems	-	-	-	-	-	-	-	-	-	-	-	2
			Avg. Cost	-	-	-	-	-	-	-	-	-	-	-	*
			Avg. Incentive	-	-	-	-	-	-	-	-	-	-	-	*
VT	RERC Small Scale Renewable	≤10 kW	No. Systems	-	-	-	-	-	-	33	15	24	60	88	134
			Avg. Cost	-	-	-	-	-	-	8.9	9.5	9.2	9.3	9.4	8.3
			Avg. Incentive	-	-	-	-	-	-	2.9	2.6	2.1	1.8	1.7	1.7

State	Program Administrator and Program Name	Size Range		1998	1999	2000	2001	2002	2003	2004	2005	2006	2007	2008	2009
	Energy Incentive Program	10-100 kW	No. Systems	-	-	-	-	-	-	-	-	-	1	5	5
			Avg. Cost	-	-	-	-	-	-	-	-	-	*	8.8	7.2
			Avg. Incentive	-	-	-	-	-	-	-	-	-	*	0.7	0.6
		>100 kW	No. Systems	-	-	-	-	-	-	-	-	-	-	-	-
			Avg. Cost	-	-	-	-	-	-	-	-	-	-	-	-
			Avg. Incentive	-	-	-	-	-	-	-	-	-	-	-	-
WI	Focus on Energy Renewable Energy Cash-Back Rewards Program	≤10 kW	No. Systems	-	-	-	-	6	9	12	22	59	86	124	225
			Avg. Cost	-	-	-	-	12.6	10.8	9.3	10.0	8.7	9.3	9.4	8.8
			Avg. Incentive	-	-	-	-	3.1	2.5	2.5	2.7	2.6	2.0	2.1	2.0
		10-100 kW	No. Systems	-	-	-	-	-	-	-	-	3	10	19	39
			Avg. Cost	-	-	-	-	-	-	-	-	*	8.2	8.5	8.6
			Avg. Incentive	-	-	-	-	-	-	-	-	*	2.2	1.9	2.0
		>100 kW	No. Systems	-	-	-	-	-	-	-	-	-	-	-	-
			Avg. Cost	-	-	-	-	-	-	-	-	-	-	-	-
			Avg. Incentive	-	-	-	-	-	-	-	-	-	-	-	-

* Average cost and incentive data are omitted if there are fewer than five systems.

** The NJ SREC-Only Pilot does not provide any direct cash incentive, but instead, provides financial support solely though the sale of solar renewable energy certificates based on solar energy production.

Appendix C: Calculating After-Tax Cash Incentives and State and Federal Investment Tax Credits

Section 4 presents trends related to combined after-tax financial incentives (direct cash incentives from state/utility PV incentive programs plus state and federal ITCs) and net installed costs after receipt of these incentives. Calculating this value required that several operations first be performed on the data provided by PV program administrators, as described below.

1) *Segmenting Systems as Residential, Commercial, or Tax-Exempt.* Data provided by many of the programs did not explicitly identify whether the PV systems were owned by residential, commercial, or tax-exempt entities. Unless otherwise identified, we classified all systems ≤10 kW as residential and all systems >10 kW as commercial.
2) *Estimating the After-Tax Value of Cash Incentives from State/Utility Incentive Programs.* Although the Internal Revenue Service (IRS) has provided only limited guidance on the issue, it appears that, in most cases, cash incentives provided for commercial PV systems are considered federally-taxable income. As such, the cash incentives provided for systems in the dataset identified as commercial PV were assumed to be taxed at a federal corporate tax rate of 35%. The taxation of cash incentives for commercial PV at the state level may vary by state; for simplicity, we assume that all commercial PV systems are taxed at the "effective" state corporate tax rate, which accounts for the fact that state corporate taxes reduce the incentive-recipient's federally-taxable income. The effective state corporate tax rate applied to the cash incentive is equal to 65% (i.e., 1 minus 35%) of the nominal state corporate tax rate in 2009, which ranged from 0% to 9.99% among the 16 states in our dataset.[52]

   Cash incentives paid to residential PV system owners are exempt from federal income taxes if the incentive is considered to be a "utility energy conservation subsidy," per Section 136 of the Internal Revenue Code. Despite several IRS private letter rulings of potential relevance, uncertainty remains as to what exactly constitutes a "utility energy conservation subsidy." Notwithstanding this uncertainty, we assume that cash incentives provided to all systems in the dataset identified as residential PV are exempt from federal income taxes. The taxation of cash incentives for residential PV at the state level may vary by state, but for simplicity, we assume that all residential PV systems are also exempt from state income tax.
3) *Estimating the Value of Federal ITCs.* Projects in the dataset identified as residential PV and installed between January 1, 2006 and December 31, 2008 were assumed to receive a federal ITC equal to the lesser of 30% of the tax credit basis or $2,000; residential PV systems installed after December 31, 2008 were assumed to receive the full 30% ITC. Projects in the dataset identified as commercial PV are assumed to receive a federal ITC equal to 10% of the tax credit basis if installed prior to January 1, 2006, or 30% of the tax credit basis if installed after that date.

Table C-1. State ITC Details

State	Applicable Customers	System Size Cap	Applicable Period	Tax Credit Amount	Cap
AZ	Residential	None	1995-indefinite	25% of *pre-rebate* installed cost	$1,000
	Non-Residential and Tax-Exempt	None	2006-2018	10% of *pre-rebate* installed cost	$25,000
CA	All	200 kW	2001-2003	15% of *net* installed cost after state rebate and federal ITC	None
	All	200 kW	2004-2005	7.5% of *net* installed cost after state rebate and federal ITC	None
MA	Residential	None	1979-indefinite	15% of *net* installed cost after federal ITC	$1,000
NY	Residential	10 kW	1998-9/1/2006	25% of *net* installed cost after rebate	$3,750
	Residential	10 kW	9/1/2006-indefinite	25% of *net* installed cost after rebate	$5,000
OR	Residential	None	11/4/2005-indefinite	$3/W based on rated capacity (DC-of STC)*	$6,000 up to 50% pre-rebate installed cost**
	Non-Residential and Tax-Exempt	None	1981-2006	35% of *pre-rebate* installed cost	$10,000,000
	Non-Residential and Tax-Exempt	None	2007-2017	50% of *pre-rebate* installed cost (up to maximum eligible cost***)	$10,000,000

*Tax credit paid out over multiple years, with an annual limit of $1,500/yr. To calculate the after-tax value of the tax credit, the stream of credits was discounted at a 10% nominal discount rate.

** For systems installed after August 13, 2010, the cap on the Oregon state tax credit for residential PV is based on net installed cost after deducting the value of any up-front cash incentives and the federal ITC. This provision, however, does not affect the systems within the data sample, all of which were installed prior to 2010.

*** The maximum eligible cost varies by system size and date of installation. The tax credit is paid out over five years. To calculate the after-tax value of the tax credit, the stream of credits was discounted at a 10% nominal discount rate.

Commercial PV systems installed in 2009 may have opted for the Treasury grant, in lieu of the ITC; for the purpose of this analysis, however, we assume that the economic value of the grant is equal to that of the ITC.

The tax credit basis on which the federal ITC is calculated depends on whether cash incentives received by a project are federally-taxable. If the cash incentives are federally-taxable, as assumed for all commercial PV, then the federal ITC is calculated based on the full installed cost of the system. If, on the other hand, the cash incentives are not federally-taxable, as assumed for all residential PV, then the federal ITC is calculated based on the installed cost minus the value of the tax-exempt cash incentives.

4) *Estimating the Value of State ITCs.* We identified five of the 16 states in our dataset as having offered a state ITC for PV at some point from 1998-2009. Based on the information contained in Table C-1, we determined whether each project in the dataset was eligible for a state ITC, and if so, estimated the amount of the tax credit. In three states (California, New York, and Massachusetts), we assumed that the basis for the state ITC was reduced for any direct cash incentives ("rebates") and/or for the federal ITC. In addition, we accounted for the fact that state tax credits are financially equivalent to federally taxable income, because they increase the recipient's federally-taxable income by an amount equal to the size of the state tax credit. The net value of state ITCs was therefore reduced by the assumed federal income tax levied on the increased income. For commercial customers, we assumed a federal income tax rate of 35%. For residential customers, we assumed that the increased income would be taxed at the marginal rate applicable to a married couple filing jointly with federally taxable income of $150,000 (e.g., 28% in 2009).[53]

Disclaimer

This document was prepared as an account of work sponsored by the United States Government. While this document is believed to contain correct information, neither the United States Government nor any agency thereof, nor The Regents of the University of California, nor any of their employees, makes any warranty, express or implied, or assumes any legal responsibility for the accuracy, completeness, or usefulness of any information, apparatus, product, or process disclosed, or represents that its use would not infringe privately owned rights. Reference herein to any specific commercial product, process, or service by its trade name, trademark, manufacturer, or otherwise, does not necessarily constitute or imply its endorsement, recommendation, or favoring by the United States Government or any agency thereof, or The Regents of the University of California. The views and opinions of authors expressed herein do not necessarily state or reflect those of the United States Government or any agency thereof, or The Regents of the University of California.

Ernest Orlando Lawrence Berkeley National Laboratory is an equal opportunity employer.

Acknowledgments

For their support of this project, the authors thank Charles Hemmeline and Jennifer DeCesaro of the U.S. DOE's Solar Energy Technologies Program and Mark Sinclair of the Clean Energy States Alliance.

For providing data or reviewing elements of this paper, the authors thank: Justin Baca and Samantha Jacoby (Solar Energy Industries Association), Mark Bolinger (Lawrence Berkeley National Laboratory), Larry Burton (Nevada Energy), Sandra Bustos (Florida Governor's Energy Office), Ron Celentano (Pennsylvania Sustainable Development Fund),

Joe Cohen (Maryland Energy Administration), Linda Crafton (Salt River Project), Libby Dodson (Pennsylvania Department of the Environment), Suzanne Elowson (Vermont Energy Investment Corporation), Steven Frantz (Sacramento Municipal Utility District), Charlie Garrison (Honeywell), John Gutenberger (Los Angeles Department of Water and Power), Tim Harvey (Austin Energy), Jeff Healion (Nevada Energy), Tom Hoff (Clean Power Research), Christie Howe (Massachusetts Clean Energy Center), Scott Hunter (New Jersey Board of Public Utilities), Leslie Libby (Austin Energy), James Lee (California Energy Commission), James Loewen (California Public Utilities Commission), Elizabeth Long (Clean Energy Associates), David Ljungquist (Connecticut Clean Energy Fund), Michael Maley (Arizona Public Service), David McLelland (Energy Trust of Oregon), Stacy Miller (Minnesota Office of Energy Security), Jeffrey Painter (Pennsylvania Department of the Environment), Jeffrey Peterson (New York State Energy Research and Development Authority), Kenneth Pritchett (Los Angeles Department of Water and Power), Bill Schutten (Wisconsin Focus on Energy), Gabrielle Stebbins (Vermont Energy Investment Corporation), Molly Sterkel (California Public Utilities Commission), Jeremy Stone (Clean Power Research), Lucy Sullivan (Connecticut Clean Energy Fund), Coryne Tasca (SENTECH), Robert Taylor (Pennsylvania Department of the Environment), Mike Winka (New Jersey Board of Public Utilities), and Steve Weise (Clean Energy Associates). We also thank Anthony Ma for assistance with cover design, formatting, and production. Of course, the authors are solely responsible for any remaining omissions or errors.

Berkeley Lab's contributions to this report were funded by the Office of Energy Efficiency and Renewable Energy (Solar Energy Technologies Program) of the U.S. Department of Energy under Contract No. DE-AC02-05CH11231, and by the Clean Energy States Alliance.

End Notes

[1] It should be noted that, in focusing specifically on *installed cost*, this report does not address performance improvements or other factors that may influence the *levelized cost of electricity* from PV.

[2] The primary data sample is based on systems funded through state and utility PV incentive programs, and therefore does not include either utility-owned or utility-scale PV systems; however, cost data from a select sample of multimegawatt systems is separately summarized within the report. Also note that, while the report is intended to portray national trends, the overall sample is heavily skewed towards systems in California and New Jersey, where the vast majority of PV systems in the United States have been installed.

[3] Unless otherwise noted, the results reflect all system types represented within the sample (e.g., rack-mounted, building-integrated, tracking, non-tracking, crystalline, thin-film, etc.).

[4] Various permutations of rating conventions may be used to describe the size of PV systems. The most common rating used by PV incentive programs is the nameplate capacity of the PV modules, which is reported by manufacturers in terms of direct current (DC) watts under standard test conditions (STC). This is the rating convention used throughout the present report.

[5] Implied non-module costs are calculated as the difference between total installed cost and Navigant's wholesale module price index. This approach cannot distinguish between actual non-module costs as paid by PV system owners and a lag between wholesale module prices and retail module costs. To characterize the long-term historical reduction in module costs, we therefore focus on the period up to 2007, as wholesale module prices were relatively flat during the preceding two years, and thus the module price lag effect in 2007 is likely to be relatively small. Discerning trends in non-module costs after 2007 using this method will require waiting for some stabilization in wholesale module pricing.

[6] Among 10-100 kW systems installed in 2009, tracking systems had average installed costs \$3.5/W higher than fixedaxis systems; however, this comparison is based on only seven tracking systems.

[7] The data provided by PV program administrators do not identify whether systems are ground-mounted or roofmounted. A more meaningful comparison, if the data were available, would be among tracking-systems, groundmounted fixed-axis systems, and roof-mounted fixed-axis systems.

[8] Throughout this report, the terms "residential PV" and "commercial PV" refer, respectively, to PV systems sited at residential and commercial customer facilities.

[9] SolarBuzz. 2010. *MarketBuzz 2010*. The report also briefly compares recent PV installed costs in the United States to those in Germany and Japan. Finally, it should be noted that the analysis presented here focuses on descriptive trends in the underlying data, serving primarily to summarize the data in tabular and http://www.solarbuzz.com/Marketbuzz2010-intro.htm.

[10] Sherwood, L. 2010. *U.S. Solar Market Trends 2009*. Interstate Renewable Energy Council. http://www.irecusa.org.

[11] To be clear, the report focuses on installed costs as paid by the system owner, rather than the costs born by manufacturers or installers. It is possible, especially over the past several years, that cost trends diverged between manufacturers and installers, or between installers and system owners.

[12] The dataset used in the present report consists of many of the same systems included in the National Renewable Energy Laboratory's OpenPV Mapping Project (http://openpv.nrel.gov). However, the results presented within this report may differ somewhat from the summary installed cost data provided through OpenPV, both because of differences in the underlying data sample, as well as potential differences in the data cleaning and processing protocols.

[13] In addition to the primary dataset, which is limited to data provided directly by PV incentive program administrators and only includes customer-sited systems, the report also summarizes installed cost data obtained through public data sources for 11 multi-MW grid-connected PV systems in the U.S. (several of which are installed on the utility-side of the meter). These additional large systems represent a combined 108 MW, bringing the total dataset to 982 MW, or 78% of all grid-connected PV capacity installed in the U.S. through 2009.

[14] All data provided to LBNL was in the form of Excel spreadsheets. A number of PV incentive program within the sample use PowerClerk, a commercial database tool specifically designed for PV incentive program tracking. We acknowledge that the use PowerClerk and any other similar products has likely increased both the quantity and quality of data available for this analysis.

[15] Various permutations of rating conventions may be used to describe the size of PV systems. The most common rating used by PV incentive programs is the total nameplate capacity of the PV modules in direct current (DC) watts under standard test conditions (STC). This is the rating convention used throughout this report. Alternatively, PV system sizes may be denominated in terms of DC watts under PVUSA test conditions (PTC), or in terms of alternating current (AC) watts under either STC or PTC.

[16] 10 kW is a common, albeit imperfect, cut-off between residential and commercial PV systems. Among the approximately 46,300 systems in the dataset for which market sector data were provided, 95% of systems (and 94% of capacity) $\leq$10 kW are residential, while 34% of systems (and 64% of capacity) >10 kW are commercial. If the same distribution applies to the entire dataset, a total of 10% of all systems in the sample (and 9% of the total capacity) would be misclassified by using 10 kW as the cut-off between residential and commercial systems.

[17] There may be a modest level of double-counting of systems between programs, as some systems funded by LADWP and SMUD may have also received incentive funding through the CEC's Emerging Renewables Program. Some other large systems funded by LADWP and SMUD also received funding through the CPUC SGIP; however, those systems were removed from the SGIP dataset, in order to eliminate double-counting.

[18] These three PV systems are the 30 MWDC DeSoto Next Generation Solar Energy Center installed in 2009 in Arcadia, FL; the 25.2 MWDC system installed in 2009 in Blythe, CA; and the 14.2 MWDC system installed in 2007 at Nellis Air Force base in Nevada.

[19] Unless otherwise noted, the reported results are based on all system types in the data sample.

[20] The rationale behind this approach is to roughly approximate a six-month lag between the time that the PV installation contract is signed and the time that the system is installed. We acknowledge, however, that this is an imperfect approximation.

[21] By comparison, the SolarBuzz retail module price index indicates that average retail module prices in the United States declined by approximately \$0.4/W from 2008 to 2009 (http://www.solarbuzz.com/Moduleprices.htm). As presented later in this report, data on component-level costs provided by a sub-set of the programs in the sample show that, from 2008-2009, average module costs declined by \$0.1/W. Reported retail module costs, however, likely include a mark-up over the wholesale module purchase price, and therefore are not strictly comparable to the Navigant Index.

[22] As with many of the trends presented in this report, small differences in averages between years or between groups of systems may, in some cases, reflect idiosyncrasies of the data, rather than fundamental market drivers.

[23] The data sample for >500 kW systems consists of ten or fewer systems each year through 2005; thus, the average costs shown for systems in this size range installed from 2003-2005 may have limited significance.

[24] Table 3 only includes systems ≥2 MW that are not in the primary dataset and for which installed cost data (or proxies thereof) could be found. Note, though, that the sources of these cost data vary in quality, and therefore these data are less certain than the data in the primary sample.

[25] Customer segment data are available for approximately 60% of the systems in the data sample. Among those systems for which customer segment data was provided, the average size of residential systems rose from 2.1 kW in 1999 to 5.5 kW in 2009, and the average size of non-residential systems rose from 4.6 kW in 1999 to 43.9 kW in 2009. did provide component-level cost data (even if at a fairly coarse level of detail),

[26] These three programs – California Solar Initiative, Minnesota's Solar Electric Rebate Program, and the Wisconsin Focus on Energy Cash-Back Rewards Program – represent 56% of the systems in the dataset installed in 2009.

[27] Installed costs may differ among countries as a result of a wide variety of factors, including differences in: module prices, technical standards for grid-connected PV systems, installation labor costs, procedures for receiving incentives and permitting/interconnection approvals (i.e., "paperwork burden"), foreign exchange rates, the degree to which components are manufactured locally, and average system size.

[28] Although Texas and Arizona are both relatively small statewide PV markets in absolute size, the data from these two states largely come from programs with relatively high local concentrations of PV system installations.

[29] For example, in Wisconsin, a relatively high-cost state, approximately 20% of the systems in the sample have tracking equipment.

[30] Customer segment identifiers were provided by PV incentive programs for approximately 92% of all 2009 installations within the dataset. We focus on the 5-10 kW, 10-100 kW, and 100-500 kW size ranges, as these are ranges for which the sample size in each relevant customer segment is sufficiently large. There were no residential systems in the 100-500 kW size range, however.

[31] The NSHP program exclusively targets residential new construction, and thus all systems in that program were assumed to be residential new construction, and all residential systems within CSI were assumed to be retrofit.

[32] Of the 1,221 NSHP systems within the dataset installed in 2009, 77% are within the 1-3 kW size range.

[33] Thin-film systems include both amorphous silicon and non-silicon modules.

[34] We limit the comparison to rack-mounted systems, in order to eliminate any distortion associated with the higher incidence of thin-film modules among BIPV systems.

[35] Although not included in the figure due to insufficient sample size, the average installed cost of the eight thin-film systems >100 kW identified within the dataset was $1.2/W higher than the crystalline systems in this size range.

[36] A more meaningful comparison would be to compare tracking systems, ground-mounted fixed-axis systems, and roof-mounted fixed-axis systems. However, the data provided by PV program administrators do not identify whether systems are ground-mounted or roof-mounted.

[37] Tracking systems increase the energy yield of the PV systems, which will tend to offset some of the increase in the up-front cost.

[38] This is the only PV incentive program that provided separate component-level cost data for the "rack" element of each PV system.

[39] Starting in 2009 and for a limited period, the federal ITC for commercial PV could be converted to a cash grant of equal value from the U.S. Treasury.

[40] For tax purposes, commercial PV owners are allowed to depreciate PV systems using an accelerated 5-year schedule. The net present value of this accelerated depreciation schedule, relative to a 20-year straight-line schedule, is equal to 12% of installed costs. See: Bolinger, M., G. Barbose, and R. Wiser. 2008. *Shaking Up the Residential PV Market: Implications of Recent Changes to the ITC*. Berkeley, CA: Lawrence Berkeley National Laboratory.

[41] PBI payments were reported by PV incentive program administrators on a $/W basis, based on estimated energy production. For the purpose of the present analysis, these reported $/W incentive amounts were discounted at a 10% nominal discount rate, based on the number of years over which the PBI payments are made.

[42] Although the IRS has provided only limited guidance on the issue, it appears that, in most cases, cash incentives provided for commercial PV systems are considered Federally-taxable income. Cash incentives for residential PV, however, are exempt from Federal income taxes if the incentive is considered to be a "utility energy conservation subsidy," per Section 136 of the Internal Revenue Code. Despite several IRS private letter rulings

of potential relevance, uncertainty remains as to what exactly constitutes a "utility energy conservation subsidy." See: Bolinger, M., G. Barbose, and R. Wiser. 2008. *Shaking Up the Residential PV Market: Implications of Recent Changes to the ITC*. Berkeley, CA: Lawrence Berkeley National Laboratory.

[43] The *American Recovery and Reinvestment Act*, signed into law in 2009, provides commercial PV projects the option, for a limited period of time, to receive an up-front cash grant from the Treasury Department of equal value to the federal ITC. For the purpose of the present analysis, we assume that the Treasury grant has the same economic value as the federal ITC for commercial PV. For a more thorough comparison of the relative value of the cash grant to the ITC, see: Bolinger, M., R. Wiser, and N. Darghouth. 2009. *Preliminary Evaluation of the Impact of the Section 1603 Treasury Grant Program on Renewable Energy Deployment in 2009*. Berkeley, CA: Lawrence Berkeley National Laboratory. LBNL-3188E.

[44] By expressing the incentives on an after-tax basis, we account for state and Federal income taxes that may be levied on direct cash incentives, as well as any reduction in the federal ITC tax basis associated with direct cash incentives. For further details, please refer to Appendix C.

[45] Two trends exhibited in Figure 21 – that aggregate after-tax incentives for commercial PV in 2009 were virtually unchanged from the prior year and were slightly greater than aggregate after-tax incentives for residential PV in 2009 –appear on fist glance to be inconsistent with Figure 20, which indicates that state/utility cash incentives declined from 2008 to 2009, and that they were generally smaller for larger systems. The reason for this apparent inconsistency is that, from 2008 to 2009, commercial PV installations in the data sample shifted disproportionately toward PV incentive programs with relatively high incentive levels.

[46] Perhaps as one sign of movement back towards the residential sector, roughly 44% of all customer-sited PV capacity installed in 2009 consisted of residential systems, compared to 32% in 2008.

[47] Wiser, R., M. Bolinger, P. Cappers, and R. Margolis. 2006. *Letting the Sun Shine on Solar Costs: An Empirical Investigation of Photovoltaic Cost Trends in California*. LBNL-59282. Berkeley, California: Lawrence Berkeley National Laboratory.

[48] See Appendix B for data on the average annual cash incentive for each of the PV incentive programs in the dataset.

[49] Within the data sample, the CORE and REIP programs represents the vast majority (99%) of New Jersey residential PV systems installed in 2009. Among commercial PV systems installed in New Jersey in 2009, however, the CORE and REIP programs represent 68% of the systems, and only 17% of the capacity, with the remaining commercial systems funded through the SREC Registration program.

[50] http://www.gosolarcalifornia.org/equipment/pvmodule.php

[51] http://www.gosolarcalifornia.org/equipment/inverter.php

[52] http://www.taxadmin.org/fta/rate/corp_inc.html

[53] http://www.taxfoundation.org/taxdata/show/151.html

In: Photovoltaic
Editors: Eric Falchuk and Carol Woodlee

ISBN: 978-161942-045-8
© 2012 Nova Science Publishers, Inc.

Chapter 3

BYPRODUCT MINERAL COMMODITIES USED FOR THE PRODUCTION OF PHOTOVOLTAIC CELLS*

Donald I. Bleiwas

INTRODUCTION

Rising fossil fuel costs, environmental concerns relating to global climate change, and Government policy to significantly increase our Nation's energy independence have placed greater emphasis on the generation of electricity from renewable sources, such as the Sun (light and heat), water, and wind, which for all intents and purposes are inexhaustible resources.

Although the total amount of electricity generated from the direct conversion of sunlight through photovoltaic cells is relatively small compared with that from other forms of renewable energy, the rate of growth in the sector is significant. The total value of energy of photovoltaic cells produced worldwide increased to nearly 7 gigawatts (GW) in 2008 from 45 megawatts (MW) in 1990, a compound annual growth rate of about 30 percent. In the United States, manufacturing of photovoltaic cells has grown exponentially to about 480 MW in 2008, accounting for 6 percent of world production, from less than 10 MW of photovoltaic capacity in 1990 (Benner, 2007; U.S. Department of Energy, Energy Information Administration, 2010), a compound annual growth rate of approximately 23 percent. A production capacity of 1 GW of electricity [or 8,760 gigawatthours[1] (GWh)] is equivalent to the annual electricity requirements for roughly 800,000 average households in the United States (U.S. Department of Energy, Energy Information Administration, 2010). This estimate does not include losses of electricity, such as during transmission through power lines.

Similarly, China has demonstrated dramatic growth in production. In 2004, China contributed about 100 MW of photovoltaic cells, about 1 percent of the global total. By 2008, China's output had grown to 1.9 GW of photovoltaic capacity, about 27 percent of global

---

* This is an edited, reformatted and augmented version of the United States Geological Survey publication, Circular 1365 dated 2010.

output (U.S. Department of Energy, Energy Information Administration, 2010), a compound annual growth rate of about 80 percent. Other major producers of photovoltaic cells in 2008 included Europe (27 percent of world production), Japan (18 percent), and Taiwan (12 percent) (U.S. Department of Energy, 2010).

It should be noted that the increase in production capacity as measured in megawatts was probably as much the result of leading electricity consuming nation in the world. The U.S. Department of Energy estimated that less than 0.02 percent of the total electricity generated in the United States in 2008 from all other renewable energy sources, including electric, wind, wood-burning, biomass, and geothermal sources, originated from photovoltaic arrays (U.S. Department of Energy, Energy Information Administration, 2010). The major reason for the low percentage was the high cost of producing electricity from photovoltaic cells compared with that from other established sources of electrical power. Wind, which is another renewable energy source often used as a comparison point against solar generation, produced 50 times the amount produced from solar sources (Blansett, 2007). For purposes of perspective, photovoltaic cells are about 7 to 17 percent efficient at converting sunlight to electricity, while electricity generated by fossil fuels (which produces about 70 percent of U.S. electricity generation) is about 28 percent efficient (U.S. Department of Energy, Energy Information Administration, 2008b).

Silicon-based photovoltaic cells currently dominate the market and are the type most often seen in solar installations and on rooftops. They account for more than 90 percent of the photovoltaic cells in use and can convert roughly 15 percent of the absorbed light to electricity in actual outdoor use. Research in recent years has resulted in increased efficiencies and lower production costs to the point where second-generation photovoltaic cells known as thin-film cells are making significant strides in penetrating the global commercial renewable energy market. The efficiency of these cells under laboratory conditions has ranged from about 20 percent to more than 40 percent, depending on composition and design (McMahon, 2008; Hecht, 2009; Novak and others, 2009).

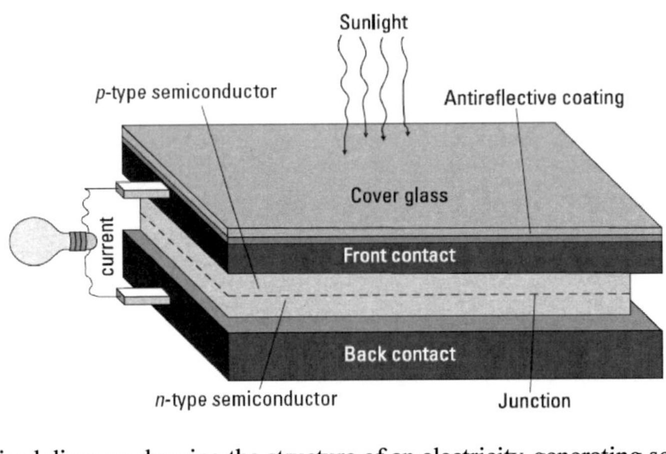

Figure 1. A generalized diagram showing the structure of an electricity-generating solar (photovoltaic) cell. Semiconductor films contain such metals as cadmium, gallium, germanium, indium, selenium, and tellurium. The p layer generates a positive charge and the n layer generates a negative charge. Front and back contacts made up of conductive metals and alloys containing aluminum, copper, gold, molybdenum, and silver. Diagram courtesy of California Energy Commission, used with permission.

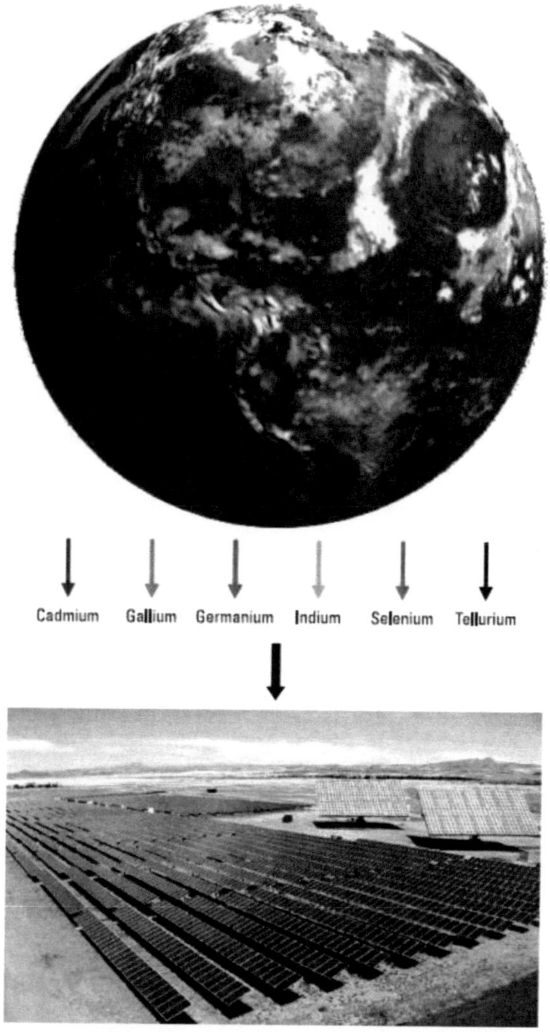

Figure 2. Metals such as cadmium, gallium, germanium, indium, selenium, and tellurium are recovered as byproducts from the production of other metals and are critical to varying extents for the efficient operation of photovoltaic cells. The current and future supply of some of these metals relies on ores mined in many countries and processed to refined metals in other countries as well as, to varying degrees, on the recovery of the metals as a byproduct of processing primary mineral commodities, fly ash produced from coal, and from recycling. Photograph of the Earth by the National Aeronautics and Space Administration (http://svs.gsfc.nasa.gov/vis/a000000/a000100/ a000155/); phorograph of solar farm by U.S. Department of Energy, National Renewable Energy Laboratory (http://www.nrel.gov/data/ pix/Jpegs/13338.jpg.

Thin-film photovoltaic cells will likely become an important source of electricity generated from sunlight in the future. Most thin-film cells have some advantages compared with their thicker amorphous silicon cell counterparts (thick-film cells). Thin-film cells are lighter because they use much less material and can be flexible, making them adaptable to various designs in different applications. The cell's active area is usually only 1 to 10 micrometers (μm) thick. Thick-film cells typically are as much as 200 to 400 μm thick (SolarThinFilms, 2007). Thin-film cells are also more easily manufactured and therefore can be mass-manufactured by automated systems and are less expensive to produce. However, when compared with crystalline silicon cells, thin-film cells are generally less efficient at converting sunlight to electricity and may require up to twice as much space to produce the same amount of electricity. However, the gap is closing with significant progress on improving the output of electricity from thin-film cells being reported from laboratory research (U.S. Department of Energy, National Renewable Energy Laboratory, 2008). Figure 1 is a generalized figure of the components that comprise a photovoltaic solar cell.

The construction and operation of facilities required for the production of electricity generated by "capturing" sunlight, specifically photovoltaic cells, require a diverse mix of mineral materials. Photovoltaic cell-based powerplants use significant tonnages of mineral materials commonly used for structural support and transmission of electricity, including aluminum, concrete, copper, glass, nickel, steel, and zinc. Land-based reserves and resources (as opposed to resources on the sea floor) that contain these materials are widely distributed over a large portion of the Earth's crust, and any supply shortages would likely be short lived when given favorable market factors. Silica, the principal raw material used for manufacturing the cells, is abundant and readily available, as is copper, another major component in some types of cells. However, some of the mineral materials currently used in more advanced thin-film photovoltaic cells incorporate some less common mineral materials that are highly valued for the special properties they impart when combined with other mineral materials. These characteristics include durability and higher levels of light absorption and conversion to electricity (Mah, 1998). These mineral materials include cadmium, gallium, germanium, indium, selenium, and tellurium (fig. 2).

Some examples of the use of these highly valued elements in thin-film photovoltaic cells include copper-indiumgallium-selenide alloy (CIGS), cadmium-tellurium (CdTe), and germanium used alone or in combination with other elements in the *p*-type (positive type) and *n*-type (negative type) silicon-based semiconductors in thin-film photovoltaic cells (fig. 1). Although these mineral commodities are used in relatively small amounts compared with other materials used in the manufacture of photovoltaic cells, they play a critical role in the performance of thin-film photovoltaic cells. With the increasing importance of photovoltaic systems as a source of supplying electricity for homes, industrial facilities, and supplemental grids in the United States and other countries, the demand for these mineral commodities will likely increase over the near term. The use of these mineral materials for other high-tech applications will also likely increase. There are concerns that the availability of mineral materials could restrain the rate of growth and the size of the industries that use these metals; however, estimates of reserves are temporal and expand and contract based on the cost of recovery and value (Wilburn and others, 2005).

This report describes the production and sources of several mineral commodities contained in thin-film photovoltaic cells that are not widely produced or are produced in relatively small quantities as well as general estimates of the amounts of the metals required to produce 1 GW. This report does not address the geographic distribution and supply of components used in the photovoltaic systems that rely on the availability of these metals. Investigation into methods of recovering materials contained in obsolete photovoltaic cells is a relatively new area of research that will likely become commercially established as the number of cells available for recycling increase. Recycling of these mineral materials is already supplementing supply from mining operations (5N Plus, 2010).

SOURCES AND SUPPLIES OF COMMODITIES USED IN PHOTOVOLTAIC CELLS

The adequacy of the supply of cadmium, gallium, germa-nium, indium, selenium, and tellurium, all of which are producedas byproducts from the production of other metals, is critically important for the implementation of large-scale global photovol-taic systems using current technologies. The supply of these commodities relies to a great extent on mining ores for the production of aluminum, coal, copper, and zinc. Considering historical prices, it would be unlikely that production of the primary products would increase in order to produce more of the six metals discussed in this report. For example, it is unlikely that more coal will be mined and burned in order to recover more germanium from ash, nor is it likely that more copper ore would be extracted to recover tellurium and more bauxite mined to recover more gallium during the production of alumina.

Supplies for most of these byproduct metals recovered from mining could potentially be increased as a result of a number of separate and related scenarios, including:

- the supply of the primary revenue-producing commodities increases as a result of an increase in the value of the metals
- new government policies are created to provide incentives for production
- byproduct metal prices increase, justifying increased production from mines and processing plants that currently produce the metals (increased production can be attained from expansion of existing facilities and construction of new circuits to recover the metals from sites where the metals are not currently produced but are potentially recoverable)
- technologies of metallurgical recovery improve
- supplies from deposits not currently producing the metals discussed can be recovered, such as new deposits discovered, past producers, and treatment of residual mine and processing waste
- the byproduct commodities are substituted with other materials and
- recycling rates of the byproduct commodities improve

Additional production might also be generated from sources where cost is not the highest priority; for example, an enterprise may be willing to contract for a guaranteed supply of metals to ensure long-term availability or a facility that is vertically integrated is able to recover the higher costs for the metals in a value-added product, such as a photovoltaic cell.

Alternatively, an increase in the price of the metal commodity, perhaps coupled with concerns about the adequacy of supply for large-scale implementation of photovoltaic technology and applications and technologies competing for the same materials, could provide incentive to investigate substitution by other materials. This in turn could produce other technological advances [such as dematerialization (using less to provide the same or greater results)] that could result in a shift in demand to other commodities or using less material with the same or improved performance or some yet undiscovered technology. More efficient cells (achieving the same or more power for less cost) could result in changes in materials use. Sudden changes in material use patterns can also occur, such as wartime materiel production of scientific breakthroughs creating revolutionary changes in materials use.

Table 1 lists the estimated world refined primary production for each of the byproduct metals discussed in this report, the primary mining industry from which most of the byproducts are recovered, and the average price for the byproduct commodity. A brief discussion of each byproduct metal shown in table 1 follows.

Cadmium

Nearly all the world's primary (recovered from mining) cadmium is derived as a byproduct of processing sphalerite, a zinc sulfide ore containing varying amounts of cadmium, indium, lead, and other recoverable or potentially recoverable metals. Zinc ores containing cadmium are mined in many countries.

The three leading producers of cadmium contained in ore in 2008, in descending order of quantity produced, were China, Peru, and Australia (Tolcin, 2010a).

Table 1. World production from primary sources and price of refined metals for byproduct commodities at yearend 2008

[e, estimated]

Byproduct commodity	Primary[1] commodities	World refinery productione (in metric tons)	Price of refined metal[e,2] (in dollars per kilogram)
Cadmium	Zinc[3,4]	16,000	6
Gallium	Aluminum, zinc[3]	111	570
Germanium	Zinc, coal[5]	140	1,490[6]
Indium	Zinc[3]	570	685
Selenium	Copper[3,7]	3,000–3,500	71
Tellurium	Copper	450–500	211

[1] Derived as a product of mining ore for the purpose of extracting the listed commodity. Secondary production refers to the recovery of metals from the recycling of materials following useful service.

[2] Rounded to three significant figures. Data are from U.S. Geological Survey (2009a,b, 2010).

[3] Prices for cadmium, indium, selenium, and tellurium are averages for 2008. Prices for gallium and germanium, are yearend 2008. All prices rounded to the nearest dollar and are reported in dollars per kilogram. Data are from U.S. Geological Survey (2009a,b, 2010). Prices may vary based on numerous factors that include contractual agreements, size of inventories, and specifications of the metals traded.

[4] An additional supply equal to about 25 percent of primary refined cadmium production (about 4,000 metric tons) originated from recycling.

[5] Recovered from fly ash, a byproduct of coal combustion for energy production.

[6] Price for polycrystalline zone-refined germanium; does not represent the unit value of entire world production. Data are from Guberman (2010).

[7] Does not include U.S. production.

Like many other commodities, the refined cadmium is smelted in countries other than the producing country. In 2008, the major producing countries of refined cadmium recovered from primary sources (and their approximated percentage shares of the world total) included China, where cadmium was refined from domestic and imported zinc concentrates (22 percent); the Republic of Korea, where virtually all cadmium was derived from imported zinc concentrates (15 percent); Japan, where cadmium was recovered from imported sources (11 percent); Kazakhstan, where cadmium was refined mostly from domestic sources (10 percent); and Canada, where cadmium was recovered from domestic, United States, and South American zinc concentrates (7 percent) (Tolcin, 2009a).

The United States has the capacity to recover primary cadmium from domestic zinc operations. In 2009, most of the Nation's zinc mines and refining facilities were on standby status because of low metal prices. Significant amounts of cadmium were not produced in 2008 owing to a reduction in zinc production worldwide. For example, cadmium production in the United States was about one-half the amount produced early in the decade. Although the actual amount is not known, a substantial amount (probably in the thousands of tons) of cadmium is not recovered worldwide and is contained in processing waste and disposed at impoundment areas because of the relatively low price of the metal.

Approximately 19,600 metric tons (t) of refined primary cadmium was produced worldwide in 2008, of which about 20 percent, or roughly 4,000 t, was produced through recycling (Tolcin, 2009a, 2010a). Total refined production of cadmium from primary and secondary sources has remained relatively steady over the decade, although the share originating from primary cadmium has probably decreased (Wilburn, 2007). The decrease in supply from mining resulted primarily from an increase in the supply of secondary cadmium recovered from recycling of obsolete nickel-cadmium (NiCd) batteries coupled with decreased use of NiCd batteries because of technological advances in batteries that use other materials, such as lithium-ion and magnesium; primary cadium production has also decreased because of overall decreased demand for metals owing to depressed global economic conditions, and concerns associated with the metal's potentially toxic effects on the environment and human health.

Approximately 20 percent of the global supply of cadmium in 2008 originated from recycled NiCd batteries, with additional amounts supplied from other materials containing cadmium. In 2008, about 80 percent of global cadmium consumption was in NiCd batteries (Tolcin, 2009a). The supply of cadmium introduced from the recycling of obsolete consumer-type rechargeable NiCd batteries has been increasing and should continue to increase for some years because NiCd batteries for many consumer uses are still favored in low-cost consumer products. However, with improved battery technologies and concerns for the environment, the amount of cadmium used for the production of NiCd type batteries is declining (Wilburn, 2007). Industrial-sized NiCd batteries are widely used and may find application in storage for electricity produced by photovoltaic systems and from other renewable forms of energy production. This use could result in reversing a decline in NiCd-battery use in consumer products in the future. Although some environmental concerns have been raised regarding the use of cadmium in photovoltaic cells, it is unclear how these concerns might ultimately affect the use of the metal. However, the issue will likely result in placing more emphasis on ensuring recycling of photovoltaic cells.

Considering the important role that zinc plays in society and the supply of cadmium recovered through recycling, it is not anticipated that the metal will become unavailable for more than short durations.

Gallium

Almost all the world's primary gallium supply originates as an impure byproduct recovered during the processing of bauxite ore to alumina, a precursor to refined aluminum. Minor amounts of gallium are also recovered from residues produced during the process to recover primary zinc metal. In 2008, approximately 111 t of primary refined gallium was produced worldwide, about 30 percent below reported capacity (Jaskula, 2010). Some of the impure gallium product is refined onsite, but much of it is sold and processed into gallium metal of differing purities elsewhere. China, Germany, and Kazakhastan are the leading producers of impure gallium. They compose roughly 50 percent of world primary production. Other major producers, in decreasing order of quantity produced, include Japan and Russia. The leading purchasers and producers of refined gallium include China, Japan, and the United States.

The amount of potentially recoverable gallium is much larger than that actually recovered owing to limited demand for the metal. Most of the gallium contained in bauxite ore is actually disposed as waste from the processing of ore to alumina. An increase in gallium demand beyond current capacity can be easily satisfied with the addition or expansion of recovery circuits at existing alumina facilities. It is unlikely that the future supply of gallium will be constrained because of limited aluminum production. Very little gallium is recovered from recycled obsolete materials.

Gallium finds its greatest application in electronic circuitry, laser diodes, and light emitting diodes (LEDs) (Kelly and Matos, 2006; Matos, 2006). Photovoltaic cells that incorporate gallium arsenide are mostly used for generating electricity for satellites and space exploration. Although gallium arsenide is well suited to photovoltaic cells, its application for consumer use is greatly limited by the high cost of producing gallium arsenide crystals, a necessary component of the cell. Lowering the cost of production will likely be key to commericialization of the technology.

Germanium

Approximately 140 t of primary germanium metal in various forms was produced in 2008, of which roughly 60 percent was recovered from residues generated during the production of refined zinc derived from ores, and about 30 percent originated from leaching fly ash, a waste product of coal combustion (David Guberman, mineral commodity specialist, U.S. Geological Survey, oral commun., April 19, 2010). The vast majority of germanium production originated from China (a mix of domestic and imported feedstock) and Canada (mostly North American in origin). Other countries with notable germanium production include Belgium (all derived from imported feedstock), Russia (mostly domestic in origin), and the United States (a mix of domestic and imported).

In 2007, germanium metal was in short supply owing to a reduction in zinc production, which was caused by low metal prices that resulted from depressed demand and increased demand for germanium for use in fiber optics and infrared technologies (Matos, 2006; Guberman, 2009, 2010). This supply shortage was reflected in the price of germanium, which increased by nearly 30 percent to $1,240 per kilogram for zone-refined germanium at yearend 2007 from about $950 per kilogram at the beginning of the year (Guberman, 2010). In 2008, the price of the metal continued to rise, and yearend 2008, was $1,490 per kilogram. Primary germanium supply relies to some extent on zinc production, which is currently operating significantly below capacity. More germanium could potentially become available with the improvement of the zinc market. The amount of germanium potentially recoverable from coal fly ash is essentially unlimited (Simons and Prinz, 1973a), but the commercial recovery of the metal is not currently viable to the extent that it has replaced germanium recovered from primary sources. The amount of germanium recovered from recycling efforts is not a significant contributor to world supply of germanium.

Indium

Approximately 570 t of indium was produced from primary sources in 2008 (Tolcin, 2010b). Most of the world's primary indium supply originates as a byproduct of processing zinc ores mined in China, Peru, and Australia (in decreasing order of quantity produced). Another source of primary indium supply is mines of other types of metals, mostly copper, lead, tin, and precious metals. As with many metals, the smelting and refining of concentrates take place in countries other than where the ore is mined. Like cadmium, most of the world's indium supply originates from zinc mining. The majority of smelting and refining of indium takes place in Canada, China, Japan, and the Republic of Korea. China recovers nearly 55 percent of the world's refined indium from mostly domestic sources. The Republic of Korea (mostly from imported concentrates), Japan (from imported concentrates), and Canada (from mostly domestic and U.S. sources) account for about 13, 11, and 8 percent of the world's primary refined indium production, respectively (Tolcin, 2009b, 2010b).

Flat-panel liquid crystal displays (LCDs) for computer monitors and flat-screen televisions are the leading application for indium (as an alloy consisting of indium tin oxide), consuming more than 50 percent of the world primary output of indium, and will likely continue to dominate the use of the element for the foreseeable future. Nearly all indium reclaimed from manufacturing of electronics, including LCDs, is also used for producing flat-panel displays (Matos, 2006; Mikolajczak, 2009). Photovoltaic cells and other competing uses may increase demand for indium. Expansions of existing recovery circuits and installations of new primary recovery circuits with improved recoveries are coming online, exploration of deposits containing anomalously high amounts of indium have become more attractive for potential development, and the recovery of indium from processing wastes (tailings and slags) is now being practiced on a limited scale and could become a significant source of future indium supply if demand justifies investment. More efficient production methods using indium tin oxide and improved technology to reclaim metals collected from the production of LCDs and other products are also being researched (Mikolajczak, 2009).

Although not currently a major source of indium, a potentially significant secondary indium industry with a focus on recycling electronic waste is likely to emerge under favorable market conditions and will potentially play a role in supplementing world indium supply as the metal finds use in other applications.

As with cadmium, the largest percentage of indium production relies on mining zinc ores. It is anticipated that the supply of indium will be able to meet anticipated industrial demand. Gallium may also potentially serve as a substitute for indium in some applications, including photovoltaic cells, if there are cost benefits (Andersson, 2000).

Selenium and Tellurium

In 2008, approximately 3,000 to 3,500 t of refined selenium and 450 to 500 t of refined tellurium were produced as a result of mining and processing ores derived from primary sources (George, 2009). Japan and Belgium are the leading producers of these metals, which are recovered from copper concentrates and residues purchased primarily from mining operations in Africa, Asia, Australia, and South America. China is also a large purchaser from these regions and also produces the metals from domestic mining operations. Although some of the selenium and tellurium metal is recovered from some gold and silver deposits with anomalously high levels of selenium and tellurium content, nearly all the world's selenium and tellurium metal produced from ore deposits depends on profitable recovery from residues (slimes) produced during the refining of copper to copper cathode; this copper cathode is derived from the mining of copper sulfide ore, from some polymetallic ore bodies (for example, the Sudbury nickel district), and from lead and zinc operations (Simons and Prinz, 1973b,c). The content of each selenium and tellurium metal in copper concentrate is generally below 100 parts per million each. The metals are commercially profitable to recover only when they are concentrated in residues collected from copper refineries and treated for the recovery of metals of value, which generally include antimony and precious metals, such as gold, palladium, platinum, and silver.

It was also estimated that, approximately 4,600 t could have been potentially recovered from copper anode slimes in 2008 (George, 2009). Mining operations that leach copper from ores do not produce anode slimes, and therefore none of the selenium or tellurium contained in the ores is recovered (Simons and Prinz, 1973b,c). Relatively small amounts of the two metals are also recovered from residues generated during the processes necessary to produce gold, lead, and zinc. Overall, roughly 50 to 60 percent of the selenium and tellurium contained in ore is refined to metal. The balance is contained in waste products, such as mill tailings, and in slag (Ojebuoboh, 2007). Because of expectations of increasing demand and higher prices for selenium metal, there is an increased focus on exploration for deposits containing selenium, which at one time was considered a deleterious component in ores, as well as tellurium. Significant improvements in recovery of selenium and tellurium at refineries will likely result in an increase in supply of selenium if prices justify recovery.

Using broad assumptions, a U.S. Department of Energy, National Renewable Energy Laboratory (2000) study estimated that approximately 900 t of copper sulfide ore extracted from copper porphyry deposits, the largest source of world primary copper, is required to recover 1 pound of byproduct tellurium and that at least 1,000 t of byproduct tellurium, or roughly ten times the world production in 2008, could potentially be produced annually from mining operations; however, for a number of reasons, chief among them being economics and proprietary technologies, the metal is not recovered at this time. This estimate is relatively close to the 1,200 t of tellurium potentially recoverable from copper anode slimes (George, 2009). An earlier study stated that approximately 0.4 pound of tellurium was contained in 2,000 pounds of blister copper, of which approximately 43 percent of the metal was potentially recoverable (Simons and Prinz, 1973c). The difference in the estimated recoverable amounts in the studies can be attributed to technological advances that have resulted in significant improvements in metallurgical recoveries since the early study was published.

The high price of tellurium and the anticipated increased demand for the metal have brought attention to a number of mineral deposits found to contain high tellurium contents that are being considered for development in China and Mexico. As is also typical with sustained higher commodity prices and (or) anticipated increased demand, efforts are concentrated more on research for improved recoveries from ores and investigations into processing the waste materials generated by producing and past mines and processing plants for potential recovery of the metals. These activities will likely result in an increase in supply of the metals when prices justify their recovery.

Only small amounts of selenium and tellurium metal are recovered by recycling obsolete materials.

Amounts of Selected Mineral Materials Required for Generating One Gigawatt of Electricity

Table 2 provides rough estimates of the amounts of the six byproduct mineral materials used in two types of thin-film photovoltaic cells sufficient to produce 8,760 GWh annually with a peak capacity of 4 GW of electricity production and an effective average daily capacity of 1 GW, based on a 25 percent capacity factor, which accounts for the hours of daylight and other factors affecting electricity production.

Depending on the type, efficiency of cells employed, and a number of other factors, a total surface area of roughly 40 square kilometers (nearly 15 square miles) of photovoltaic panels could be required to produce this amount of electricity assuming a peak wattage output of 110 watts per square meter. The area required to accommodate other necessary parts of generating plants is not included in this estimate. An area of roughly 33,000 square kilometers (13,000 square miles) might be required to meet the electricity needs of the United States using only photovoltaic cells (New Mexico Solar Energy Association, 1999). This is not to suggest that a single plant would be required to occupy this amount of area for the production of the electrical energy; rather it would likely require a significant number of plants distributed throughout the United States. For perspective, the largest operating photovoltaic plant in the world is the Parque Fotovoltaico Olmedilla de Alarcon solar farm in Spain which is designed to produce about 60 MW at peak levels. In California, the proposed 550MW Topaz Solar Farm Project is planned to produce nearly 1.1 GWh using thin-film CdTe photovoltaic cells when it becomes fully operational in 2013 (Pacific Gas and Electric Company, 2008). If the plant's construction follows the original design, the plant will be the world's largest, encompassing an area of about 25 square kilometers (9.5 square miles) (Hudson, 2008; Rigley, 2009).

Table 2. Metals required[1] to produce thin-film photovoltaic cells with effective annual capacity of 8,760 gigawatthours

[CIGS, copper-indium-gallium-selenide alloy; e, estimated; XX, not applicable]

Type of photovoltaic technology	Metals required			
	Metal	Quantity[e,2] (metric tons)	Percentage of 2008 estimated world refinery production from primary sources[3]	Value of contained metal[e,4] (in millions of dollars)
Thin-film CIGS	Gallium	30	27	17
	Indium	90	16	62
	Selenium	180	6	13
	Total	XX	XX	92
Thin-film cadmium telluride	Cadmium	340	2	2
	Tellurium	390	82	82
	Total	XX	XX	84

[1] Metals required to produce photovoltaic cells that can generate 4 gigawatts (GW) of peak power or 8,760 gigawatthours per year (GWh/yr) of effective capacity. One GW is equivalent to 1 billion watts, 1,000 megawatts, and 1 million kilowatts of effective capacity of electricity production. Applying a 25 percent capacity factor, an installed peak capacity of approximately 4 GW would be required to produce 1 GW of electricity on an average daily basis or 8,760 GWh/yr, electricity sufficient to meet the average annual need of 11,000 kilowatthours for 800,000 households in the United States (U.S. Department of Energy, Energy Information Administration, 2009a). No energy storage, such as batteries, was assumed.

[2] Calculation for metal requirements assumed to be linear and based on a 3-micrometer (μm) thickness cell. Two-μm thickness cells are in development and are expected to penetrate the commercial market within the next several years. The combination, amounts, and efficiency of materials used in photovoltaic cells can vary; the amounts and combinations listed in this table are not representative of all technologies. Estimates for material requirements should be considered with caution. The amount of material required (the size of the array) to produce a given amount of energy can vary greatly by the physical location of photovoltaic arrays that affect insolation (such as latitude and weather conditions) and design factors (such as the type of cells and whether the system is fixed tilt or tracking). Estimates are based on a 25 percent capacity factor. Values are rounded to two significant figures.

[3] Rounded to two significant figures.

[4] Rounded to two significant figures. Prices for cadmium, indium, selenium, and tellurium are average 2008; prices for gallium and germanium are yearend 2008. All prices are rounded to the nearest dollar and are reported in dollars per kilogram. Data are from U.S. Geological Survey (2009a,b, 2010). Prices may vary based on numerous factors that include contractual agreements, size of inventories, and specifications of the metals traded.

The estimates presented in table 2 provide a perspective for selected material requirements in three types of photovoltaic cells used mostly for commercial purposes. Other types of photovoltaic cells, such as gallium-arsenide- and germanium-based cells, are mainly reserved for use in satellites and space exploration because of their high efficiencies. They are much more expensive compared to most other types of cells in use and not economically practical for large-scale commercial application, but research is ongoing. The estimates in table 2 should be considered to be a "snapshot" because advances in the field of photovoltaic technology that require different amounts and types of materials are announced frequently. Actual material requirements could differ significantly from the estimates because cell types are generic categories and do not have a standardized composition. Within cell types, the amounts and selections of materials used and their structure vary. In addition, information about characteristics of cells is usually considered proprietary and therefore is difficult to acquire.

The amount of electricity produced from photovoltaic cells of the same composition and type will differ in practice because of numerous criteria that include the design of the array (fixed-position or tracking), geographic location, including altitude and latitude (determines the number of hours and the angle of sunlight), and prevailing climatic conditions, primarily cloud cover during daylight hours. These factors result in an electricity production efficiency factor that can vary from 15 to 35 percent of production under optimum (peak) conditions. The calculations in table 2 use an assumed capacity factor of 25 percent (estimates range from about 20 to 30 percent for the 48 contiguous United States). For these reasons, material requirements and electrical power production are most often standardized under ideal circumstances and as peak power per panel or as electricity produced by photovoltaic panel surface area at peak power. A capacity factor is applied for planning and construction purposes.

SUMMARY

Metals such as cadmium, gallium, germanium, indium, selenium, and tellurium are important mineral materials used in current photovoltaic cell technology. Most of the world's primary supply of these mineral commodities is recovered as byproducts from ores processed mainly for the purpose of producing aluminum, copper, lead, and zinc and, in the case of germanium, from ash derived from burning coal for the production of energy. In some cases, the refined metals originate in the countries in which they are mined, while in other cases, intermediate products containing the metals and refined metals are recovered by processers located in other countries. Assuming that the modest rate of market penetration of electricity from photovoltaic cells and competing markets increases demand for cadmium, gallium, and germanium, the resulting new demand can be met in relatively quick order with the expansion of existing recovery circuits or the addition of new circuits. It is very unlikely that, under most scenarios, the need would arise to mine more bauxite, copper, and zinc ore or to burn more coal to satisfy anticipated material requirements to meet the photovoltaic-generated demand for these mineral materials. For other metals, such as indium, selenium, and tellurium, targeted mineral exploration and improvements in metallurgical recoveries are likely to be needed to meet demand requirements. Substitution for these materials in nonphotovoltaic applications could also "free up" supply. Conversely, however, other new applications can produce supply constraints or increase metal prices and place pressure on their use in photovoltaic technology.

The supply of these metals from secondary sources will likely increase over time when photovoltaic cells are recycled as a result of breakage or when they reach the end of their useful life of approximately 20 to perhaps 30 years (U.S. Department of Energy, Solar Technologies Program, 2008a).

Other than perhaps short term interruptions resulting from market forces or geopolitical events, it is not anticipated that there will be any long term material constraints that would prevent the development of a significant amount of energy from photoelectric cells. Reserve estimates are not static; although a particular metal might be considered rare and scarce, if a profit can be made from its recovery, then there will likely be no long-term shortage. Technological advancements driven by the desire to produce energy at a lower cost will likely result in increases in efficiency requiring smaller amounts of these metals per unit of energy produced, substitution with other materials, and other advancements in science.

References

5N Plus, 2010, 5N Plus—Pure and simple: 5N Plus, accessed April 7, 2010, at http://www.5nplus.com/index.php/en/.

Andersson, B.A., 2000, Materials availability for large-scale thin-film photovoltaics; Progress in photovoltaics: Research and Applications, v. 8, no. 1, p. 61–76, accessed May 7, 2009, at http://www3.interscience. wiley.com/ journal/70001629/abstract.

Benner, John, 2007, Materials demand for the rapidly expanding solar electricity industry: U.S. Department of Energy, National Renewable Energy Laboratory, March 7, accessed May 5, 2009, at http://cdnintlminerals.com/i/ pdf/3Party/2007-03-07_NREL.pdf.

Blansett, Christopher, 2007, Solar energy; Solar finance and forecast conference: J.P. Morgan, accessed May 15, 2009 at http://www. jpmorgan.com/cm/Satellite?blobcol=urldata&b lobheader= application% 2Fpdf&blobkey=id&blobtable=MungoBlobs&blobwhere = 1158480 318554&ssbinary=true&b lobheadername1=Content-disposition &blob headervalue1= attachment;filename=Solar_ Energy:__Solar_ Finance_ and_ Forecast_Conference.pdf.

George, M.W., 2009, Selenium and tellurium, in Metals and minerals: U.S. Geological Survey Minerals Yearbook 2008, v. I, p. 65.1–65.9, accessed May 15, 2009, at http://minerals.usgs.gov/minerals/pubs/commodity selenium/myb1-2008-selen.pdf.

Guberman, D.E., 2009, Germanium, in Metals and minerals: U.S. Geological Survey Minerals Yearbook 2008, v. I, p. 30.1–30.7, accessed May 15, 2009, at http://minerals.usgs.gov/minerals/pubs/commodity germanium/ myb1-2008-germa.pdf.

Guberman, D.E., 2010, Germanium, in U.S. Geological Survey Mineral Commodity Summaries 2010: U.S. Geological Survey, p. 64–65, accessed June 7, 2010, at http://minerals.usgs.gov/minerals/pubs/mcs/.

Hecht, Jeff, 2009, Photonic frontiers—Thin-film photovoltaics— The key to success is higher efficiency: OptoIQ, v. 45, no. 3, March 1, accessed June 5, 2009, at http://www.optoiq.com/ index/display/article-display.articles. optoiq2.photonics-technologies.applications-__markets.energy-__environment. solar-photovoltaics.2009.12.photonic-frontiers-thinfilm-photovoltaics-the-key-to-success-is-higher-efficiency. QP129867.dcmp=rss.page=1.html.

Hudson, Gavin, 2008, World's 13 biggest solar energy plants, photovoltaic: Ecolocalizer, March 5, accessed February 24, 2009, at http://ecoworldly.com/2008/03/05/ worlds-7-biggest-solar-energy-plants/.

Jaskula, B.W., 2010, Gallium, in U.S. Geological Survey Mineral Commodity Summaries 2010: U.S. Geological Survey, p. 58–59, accessed June 7, 2010, at http://minerals.usgs.gov/minerals/pubs/mcs/.

Kelly, T.D., and Matos, G.R., 2006, Historical statistics for mineral and material commodities in the United

States: U.S. Geological Survey Data Series 140, accessed April 12, 2010, at http://minerals.usgs.gov/ds/2005/140/.

Mah, Olivia, 1998, Fundamentals of photovoltaic materials: National Solar Power Research Institute, Inc., 10 p., accessed May 8, 2009 at http://userwww.sfsu.edu/~ciotola/solar/pv.pdf.

Mikolajczak, Claire, 2009, Availability of indium and gallium—September 2009: Indium Corporation, 10 p., accessed April 8, 2010, at http://www.indium.com/_dynamo/download.php?docid=552.

Matos, Grecia, 2006, Effects of regulation and technology on end uses of nonfuel mineral commodities in the United States: U.S. Geological Survey Scientific Investigations Report 2006–5194, accessed April 12, 2010, at http://pubs.usgs.gov/sir/2006/5194/.

McMahon, Síle, 2008, NREL boosts CIGS thin-film efficiency to record 19.9 percent: Semiconductor Media Limited PV-tech.org, March 25, accessed June 5, 2009, at http://www.pv-tech.org/news/_a/nrel_boosts_ cigs_thin_ film_solar_cell_efficiency_to_record_199_percent/.

New Mexico Solar Energy Association, 1999, The solar resource: New Mexico Solar Energy Association, accessed May 7, 2009, at http://joewerne.com/eos/text/nmsea.html.

Novak, Erik, Hopkins, Stephen, and Masters, Andrew, 2009, Optical surface profiling—3-D surface characterization improves solar cell efficiency: OptoIQ, December 23, accessed January 4, 2010, at http://www.optoiq.com/index/photonics- optoiq2/photonics - technologies/applications-__markets/ energy-__environment/solar-photovoltaics/2009/12/ optical-surface-profiling-3-d-surface-characteri zationimproves - solar-cell-efficiency.html.

Ojebuoboh, Funsho, 2007, Selenium and tellurium from copper refinery slimes and their changing applications: European Metallurgical Conference, June 11–14, 2007, Düsseldorf, Germany, Proceedings, v. 1, p. 571–585, accessed May 5, 2009, at http://www.emc.gdmb.de/2007/ proceedings 2007cont.pdf.

Pacific Gas and Electric Company, 2008, PG&E signs historic 800 MW photovoltaic solar power agreements with OptiSolar and SunPower: Pacific Gas and Electroic Company news release, August 14, accessed June 7, 2010, at http://www.pge.com/about/news/mediarelations/ newsreleases/q3_2008/080814.shtml.

Rigley, Colin, 2009, It's kick-off time for solar: New Times, v. 23, no. 44, accessed May 12, 2009, at http://www.newtimesslo.com/news/2689/ its-kickoff-time-for-solar/.

Simons, F.S., and Prinz, W.C., 1973a, Gallium, germanium, and indium, in Probst, D.A., and Pratt, W.P., eds.,

United States mineral resources: U.S. Geological Survey Professional Paper 820, p. 237–250, accessed June 7, 2010, at http://pubs.er.usgs.gov /usgspubs/pp/pp820/.

Simons, F.S., and Prinz, W.C., 1973b, Selenium, in Probst, D.A., and Pratt, W.P., eds., United States mineral resources: U.S. Geological Survey Professional Paper 820, p. 573–576, accessed June 7, 2010, at http://pubs.er.usgs.gov/usgspubs/ pp/pp820/.

Simons, F.S., and Prinz, W.C., 1973c, Tellurium, in Probst, D.A., and Pratt, W.P., eds., United States mineral resources: U.S. Geological Survey Professional Paper 820, p. 627–630, accessed June 7, 2010, at http://pubs.er.usgs.gov/usgspubs/ pp/pp820/.

SolarThinFilms Inc., 2007, Thin-film photovoltaics: SolarThinFilms Inc., accessed May 21, 2009, at http://www.solarthinfilms.com/ active/en/home/ photovoltaics/ photovoltaics_and_thinfilms/ thinfilm_ photovoltaics.html.

Tolcin, A.C., 2009a, Cadmium, in Metals and minerals: U.S. Geological Survey Minerals Yearbook 2008, v. I, p. 15.1–15.8, accessed May 15, 2009, at http://minerals.usgs.gov/minerals/pubs/commodity cadmium/ myb1-2008-cadmi.pdf.

Tolcin, A.C., 2009b, Indium, in Metals and minerals: U.S. Geological Survey Minerals Yearbook 2008, v. I, p. 35.1–35.8, accessed May 15, 2009, at http://minerals.usgs.gov/minerals/pubs/commodity myb1-2008-indiu.pdf.

Tolcin, A.C., 2010a, Cadmium, in U.S. Geological Survey Mineral Commodity Summaries 2010: U.S. Geological Survey, p. 36–37, accessed June 7, 2010, at http://minerals.usgs.gov/minerals/pubs/mcs/.

Tolcin, A.C., 2010b, Indium, in U.S. Geological Survey Mineral Commodity Summaries 2010: U.S. Geological Survey, p. 74–75, accessed June 7, 2010, at http://minerals.usgs.gov/minerals/pubs/mcs/.

U.S. Department of Energy, 2010, 2008 solar technologies market report: U.S. Department of Energy, National Renewable Energy Laboratory DOE/GO–102010–2867, 119 p., accessed June 12, 2009, at http://www1.eere. energy.gov/ solar/pdfs/46025.pdf.

U.S. Department of Energy, Energy Information Administration, 2009a, Frequently asked questions—Electricity: U.S. Department of Energy, Energy Information Administration, accessed February 22, 2010, at http://tonto.eia.doe.gov/ask/ electricity_faqs.asp#electricity_use_home.

U.S. Department of Energy, Energy Information Administration, 2009b, Electricity—U.S. data: U.S. Department of Energy, Energy Information Administration, accessed May 5, 2009, at http://www.eia.doe.gov /fuelelectric.html.

U.S. Department of Energy, Energy Information Administration, 2010, Independent statistics and analysis, Net generation by energy source—Total (all sectors): U.S. Department of Energy, Energy Information Administration, accessed April 22, 2010, at http://www.eia.doe.gov /cneaf/electricity/ epm/table1_1.html.

U.S. Department of Energy, National Renewable Energy Laboratory, 2000, Assessment of critical thin film resources— Tellurium: National Renewable Energy Laboratory Contract no. RAF–9–29609, 34 p., accessed May 7, 2009, at http://www.nrel.gov/pv/thin_ film/docs/ telluriumworldindustrialminerals2000.doc.

U.S. Department of Energy, National Renewable Energy Laboratory, 2008, Record makes thin-film solar cell competitive with silicon efficiency: U.S. Department of Energy, National Renewable Energy Laboratory press release NR–0408, accessed June 2, 2009, at http://www.nrel.gov/ news/press/2008/574.html.

U.S. Department of Energy, Solar Technologies Program, 2008, Solar FAQs—Photovoltaics—All: U.S. Department of Energy, accessed April 1, 2009 at http://apps1.eere.energy name=Photovoltaics/cat=ALL.

U.S. Geological Survey, 2009a, Metals and minerals: U.S. Geological Survey Minerals Yearbook 2008, v. I, accessed April 12, 2009, at http://minerals.usgs.gov/minerals/pubs/ commodity/myb/.

U.S. Geological Survey, 2009b, Mineral commodity summaries 2009: U.S. Geological Survey, 195 p., accessed May 14, 2009, at http://minerals.usgs.gov/minerals/pubs/ mcs/.

U.S. Geological Survey, 2010, Mineral commodity summaries 2010: U.S. Geological Survey, 195 p., accessed April 14, 2010, at http://minerals.usgs.gov/minerals/pubs/mcs/.

Wilburn, D.R., 2007, Flow of cadmium from rechargeable batteries in the United States, 1996–2007: U.S. Geological Survey Scientific Investigations Report 2007–5198, 26 p., accessed February 4, 2010, at http://pubs.usgs.gov/ sir/2007/5198/.

Wilburn, D.R., Goonan, T.G., and Bleiwas, D.I., 2005, Technological advancement—A factor in increasing resource use: U.S. Geological Survey Open-File Report 2001–197, ver. 1.03, 81 p., accessed February 4, 2010, at http://pubs.usgs.gov/of/2001/of01-197/2001-197.pdf.

End Note

[1] The amount of power generated during the course of 1 year.

INDEX

D

E

F

G

H

I

J

L

M

Y

Z